ANATOMY
RECALL

ANATOMY RECALL

**RECALL SERIES EDITOR
AND SENIOR EDITOR
LORNE H. BLACKBOURNE, M.D.**
General Surgeon
Fayetteville, North Carolina

**EDITORS
JARED ANTEVIL, M.D.**
United States Marine Corps
Camp Pendleton, California
CHRISTOPHER MOORE, M.D.
Resident in Emergency Medicine
Carolinas Medical Center
Charlotte, North Carolina

LIPPINCOTT WILLIAMS & WILKINS
A **Wolters Kluwer** Company
Philadelphia · Baltimore · New York · London
Buenos Aires · Hong Kong · Sydney · Tokyo

Acquisitions Editor: Elizabeth A. Nieginski
Editorial Director: Julie P. Martinez
Development Editor: Melanie Cann
Managing Editor: Amy Dinkel
Marketing Manager: Aimee Sirmon

9 8 7 6 5 4

Care has been taken to confirm the accuracy of the information presented and to describe generally accepted practices. However, the authors, editors, and publisher are not responsible for errors or omissions or for any consequences from application of the information in this book and make no warranty, express or implied, with respect to the contents of the publication.

The authors, editors, and publisher have exerted every effort to ensure that drug selection and dosage set forth in this text are in accordance with current recommendations and practice at the time of publication. However, in view of ongoing research, changes in government regulations, and the constant flow of information relating to drug therapy and drug reactions, the reader is urged to check the package insert for each drug for any change in indications and dosage and for added warnings and precautions. This is particularly important when the recommended agent is a new or infrequently employed drug.

Some drugs and medical devices presented in this publication have Food and Drug Administration (FDA) clearance for limited use in restricted research settings. It is the responsibility of the health care provider to ascertain the FDA status of each drug or device planned for use in their clinical practice.

CONTRIBUTORS

Wang Cheung, M.D.
Jamal Hairston, M.D.
Meredith LeMasters, M.D.
Steven Liu, M.D.
Bruce Lo, M.D.
Anu Meura, M.D.
Suzanne Perks, M.D.
Andrew Wang, M.D.
Thomas Wang, M.D.
Philip Zapata, M.D.

Dedication

This book is dedicated to the medical students at the University of Virginia.

Dedication

Contents

Preface

Anatomy Recall was written by medical students, physicians, and anatomists specifically for use during a first-year gross anatomy course and as a review for the United States Medical Licensing Examination (USMLE) Step I. While there are certainly a wealth of gross anatomy texts available, most are better suited for reference than for mastery of the basic anatomy required to be a successful medical student and physician. It is our intention that *Anatomy Recall* and an atlas are all you will need for a comprehensive study of basic anatomy.

Anatomy Recall is arranged in the extremely successful question-and-answer format that defines the entire *Recall* series—a format that emphasizes active acquisition of knowledge, rather than passive absorption of it. Where appropriate, simple figures have been included to supplement the text material. Each chapter concludes with a "power review" that covers the most important and frequently tested facts in each subject area. These power reviews are ideal for a quick review prior to an anatomy examination, a board examination, or a surgery clerkship.

Anatomy is an exciting yet demanding course. It is important to have a text that is comprehensive yet readable and emphasizes (and reemphasizes) key points. A thorough initial study of anatomy will continue to reward you throughout a lifetime of clinical practice. It is our hope that *Anatomy Recall* will prove to be an invaluable tool for mastering the subject of anatomy. Good luck!

The Editors

Acknowledgments

The editors would like to acknowledge Melanie Cann, Amy Dinkel, Julie Martinez, and Elizabeth Nieginski at Lippincott Williams & Wilkins for their help and vision in bringing this book to fruition.

1 Overview

It is important to adhere to a certain formalism when describing the location or movement of one body part relative to another; therefore, a significant portion of the anatomy course (like many introductory courses in medicine) is devoted to teaching a language necessary for communicating with other healthcare professionals.

ANATOMIC POSITION

What standard position is assumed when describing the human body?

That of a human standing facing forward, feet pointing forward and palms facing outward (the "anatomic position")

ANATOMIC PLANES

Describe the three basic anatomic planes.

1. **Transverse (horizontal):** A horizontal plane across the body in anatomic position; the most common cut used in computed tomography (CT) and magnetic resonance imaging (MRI)
2. **Sagittal:** A plane formed by a vertical midline cut that divides the body into right and left sides
3. **Coronal (frontal):** A plane formed by a cut across the body in anatomic position from side to side and top to bottom

ANATOMIC DESCRIPTORS

Define the following terms:

Ventral

Toward the anterior (or front) of the body

Dorsal

Toward the posterior (or back) of the body

Medial

Closer to the midline

Lateral

Further from the midline

In the anatomic position, are the thumbs medial or lateral to the forefinger?

With the palms facing up, the thumbs are lateral to the other fingers.

What is the position of the great toe (first toe) relative to the other toes?

The great toe is medial.

Define the following terms:

    Proximal

Closer to the center of the body (often considered the heart)

    Distal

Further from the center of the body

Where is the radial artery in relation to the subclavian artery?

The radial artery (in the forearm) is distal to the subclavian artery (under the clavicle)

Which is more distal, the femur or tibia?

The tibia

ANATOMIC MOVEMENTS

What are the three major types of muscle?

Smooth, cardiac, and skeletal (striated)

Describe the innervation and characteristics of skeletal muscle.

Skeletal muscle is generally innervated by somatic nerves (i.e., movement is voluntary), and is located between two stable points (i.e., bones). Contraction results in movement of a structure.

What four parameters are used to describe skeletal muscles?

Origin: Usually the more proximal, more medial, and more stable structure that the muscle is attached to
Insertion: Usually the more distal, more lateral structure that the muscle is attached to, and the one that is moved by contraction
Innervation: The nerve that causes the muscle to contract
Action: The result of the muscle contracting

Define the following muscle actions:

    Flexion

Decreasing the angle of a joint, or bending the joint

Extension	Increasing the angle of the joint, or straightening the joint
Abduction	Moving one structure away from another laterally (i.e., away from anatomic position)
Adduction	Moving one structure toward another medially (i.e., toward the anatomic position)—think *add = together*

Describe the action that occurs with each of the following movements:

Kicking a soccer ball	Extension of the leg at the knee
Spreading the fingers	Abduction of the fingers at the metacarpophalangeal joints
Bringing an arm that is extended straight out and to the side laterally, toward the body	Adduction of the arm at the shoulder

What is the difference between ligaments and tendons?	Tendons attach the muscle to the bone, while ligaments attach bone to bone.
What is a strain?	A partial or incomplete tear of a muscle or ligament
What is a sprain?	A partial or incomplete tear of a tendon

2   The Head

THE SKULL

What is the skull?
The skeleton of the head, including the mandible

What are the two regions of the skull?
The neurocranium (i.e., the portion of the skull that encloses the brain) and the facial cranium

What is the calvaria?
The skull cap (i.e., the vault of the neuro-cranium, or the portion of the skull that is left when the facial bones are removed)

What is diploë?
The spongy bone layer between the dense outer and inner bone layers of the calvaria

Identify the structures on the following lateral view of the skull:

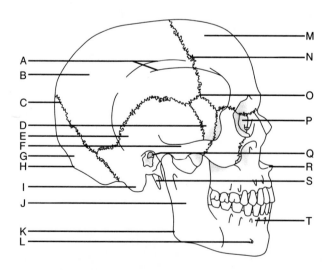

A = Inferior and superior temporal lines
B = Parietal bone
C = Lambdoid suture
D = Sphenoid bone, greater wing
E = Temporal bone
F = Zygomatic arch
G = Occipital bone
H = External occipital protuberance
I = Mastoid process
J = Ramus of the mandible
K = Angle of the mandible
L = Mental foramen
M = Frontal bone
N = Coronal suture
O = Pterion (the "p" is silent)
P = Lacrimal bone
Q = External auditory (acoustic) meatus
R = Anterior nasal spine
S = Styloid process
T = Alveolar process

What are the superior and inferior temporal lines?

The attachment points for the temporalis muscle

What region lies below the superior and inferior temporal lines?

The temporal fossa

What is the clinical significance of the proximity of the external auditory meatus and the mastoid process?

Severe middle ear infections may spread to the mastoid process of the temporal bone.

NEUROCRANIUM

Bones and sutures

Which eight bones make up the neurocranium?

The frontal bone, the two parietal bones, the two temporal bones, the occipital bone, the sphenoid bone, and the ethmoid bone

What are the immobile junctions between the bones of the neurocranium called?

Sutures

Which bones articulate at the:

    Coronal suture?   The frontal and parietal bones

    Sagittal suture?   The parietal bones of either side

    Lambdoid suture?   The parietal and occipital bones

What is the intersection of the lambdoid and sagittal sutures called?   The lambda

What is the intersection of the sagittal and coronal sutures called?   The bregma

What is a metopic suture?   A persistent frontal suture, present in approximately 2% of the population

What is craniosynostosis?   Premature closure of the sutures

What is:

    Scaphocephaly?   Premature closure of the sagittal suture

    Acrocephaly?   Premature closure of the coronal suture

    Plagiocephaly?   Premature closure of the coronal and lambdoid sutures on one side only

Identify the labeled points on the neurocranium on the following posterior and superior views:

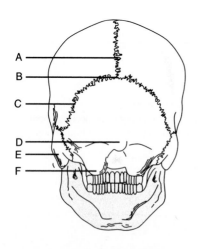
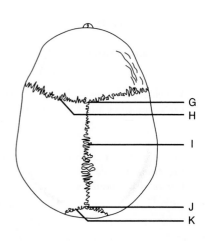

A = Sagittal suture
B = Lambda
C = Lambdoid suture
D = External occipital protuberance
E = Mastoid process
F = Occipital condyle
G = Bregma
H = Coronal suture
I = Sagittal suture
J = Lambda
K = Lambdoid suture

What are fontanelles?

Large fibrous areas where several sutures meet; often called "soft spots" on an infant's head

What are the largest fontanelles, and where are they located?

The anterior and posterior fontanelles, on the superior surface of the neurocranium

Which sutures form the borders of the posterior fontanelle?

The sagittal and lambdoid sutures

How can the anterior and posterior fontanelles be identified on an infant?

The anterior fontanelle is diamond-shaped and palpable in children younger than approximately 18 months. The posterior fontanelle is triangular and is not palpable past 1 year of age.

In adults, what is the name of the remnant of the:

Anterior fontanelle?

The bregma

Posterior fontanelle?

The asterion

What is the location of the anterolateral (sphenoidal) fontanelle called in adults?

The pterion (brain surgery using an anterolateral incision is called a "pterional approach")

Why is the pterion clinically significant?

The thinnest part of the lateral skull, the pterion is vulnerable to fractures that can damage the middle meningeal artery, which lies on the internal skull surface in this region.

Internal surface features

Label the following view of the floor of the neuro-cranium:

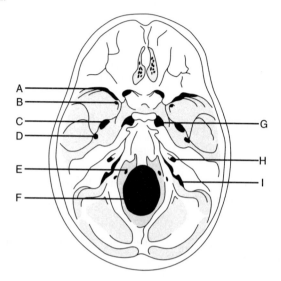

A = Superior orbital fissure
B = Foramen rotundum
C = Foramen ovale
D = Foramen spinosum
E = Hypoglossal canal
F = Foramen magnum
G = Foramen lacerum
H = Internal auditory meatus
I = Jugular foramen

Anterior cranial fossa

In addition to the ethmoid bone, which bone contributes to the floor of the anterior fossa?	The frontal bone
What is the name of the flat part of the ethmoid bone that lies anteriorly in the midline?	The cribriform plate
What structure passes through the cribriform plate?	Cranial nerve (CN) I (the olfactory nerve)

What is the name of the sharp upward projection of the ethmoid bone in the midline?

The crista galli

What is the function of the crista galli?

It provides the anterior attachment site for the falx cerebri (i.e., the dural fold that lies in the longitudinal fissure between the two cerebral hemispheres).

Which structures pass through the anterior and posterior ethmoidal foramina?

The anterior and posterior ethmoidal nerves and vessels, respectively

Middle cranial fossa

Which part of the brain occupies the middle cranial fossa?

The temporal lobes of the cerebral hemispheres

What are the borders of the middle cranial fossa:

Anteriorly?

The lesser wings of the sphenoid bones

Posteriorly?

The petrous part of the temporal bone

Laterally?

The squamous part of the temporal bone, the greater wings of the sphenoid bones, and the parietal bones

Ventrally?

The temporal bones and the greater wings of the sphenoid bones

Which three structures pass from the middle cranial fossa into the orbit via the optic canal?

1. CN II (the optic nerve)
2. The ophthalmic artery (a branch of the internal carotid artery)
3. The central vein of the retina

Which opening between the greater and lesser wings of the sphenoid bone connects the middle cranial fossa with the orbit?

The superior orbital fissure

Which six structures pass from the middle cranial fossa to the orbit through the superior orbital fissure?

1. CN III (the oculomotor nerve)
2. CN IV (the trochlear nerve)
3. CN V_1 (the ophthalmic division of the trigeminal nerve)
4. CN VI (the abducens nerve)
5. The superior ophthalmic vein
6. The inferior opthalmic vein

The foramen rotundum transmits structures between which two spaces?

The middle cranial fossa and the pterygopalatine fossa

Which structure passes through the foramen rotundum?

CN V_2 (the maxillary division of the trigeminal nerve)

The foramen ovale transmits structures between which two spaces?

The middle cranial fossa and the infratemporal fossa

Which two structures pass through the foramen ovale?

CN V_3 (the mandibular division of the trigeminal nerve) and the accessory meningeal artery

The foramen spinosum connects the middle cranial fossa with which space?

The infratemporal fossa (like the foramen ovale)

Which structure passes through the foramen spinosum?

The middle meningeal artery

The foramen lacerum lies at the junction of which cranial bones?

The sphenoid bone and the petrous part of the temporal bone

Grooves on the anterior part of the petrous temporal bone transmit which structures?

The greater and lesser petrosal nerves

What is the name of the thin plate of bone located at the junction of the petrous and squamous parts of the temporal bone?

The tegmen tympani

What is the clinical significance of this thin bone?

This bone, which separates the tympanic cavity from the middle cranial fossa, is so thin that infections of the middle ear can spread to the meninges and brain.

What is the name of the elevation of the sphenoid bone between the two optic canals?

The tuberculum sellae

What is the name of the depression posterior to the tuberculum sellae?

The sella turcica ("Turkish saddle")

What is the name of the bony ridge that defines the posterior limit of the sella turcica?

The dorsum sellae

What are the boundaries of the sella turcica:

Anteriorly?

The tuberculum sellae

Posteriorly?

The dorsum sellae

Which structure lies in the hypophyseal fossa of the sella turcica?

The pituitary gland

Which space is located directly inferior to the sella turcica?

The sphenoid sinus (surgery on the pituitary gland uses a "trans-sphenoidal" approach)

Which structure forms the roof of the sella turcica?

The diaphragma sellae (i.e., one of the dural folds)

Which processes project from the lateral aspects of the dorsum sellae?

The posterior clinoid processes

What structures attach to the posterior clinoid processes?

The tentorium cerebelli (i.e., the dural fold between the occipital lobes and the cerebellum)

Posterior cranial fossa

Which part of the brain lies in the posterior cranial fossa?

The cerebellum and brain stem

What are the borders of the posterior cranial fossa:

Anteriorly?

The petrous part of the temporal bone

Posteriorly?

The occipital bone

Ventrally (i.e., the floor)?

The occipital bone and the mastoid processes of the temporal bones

Dorsally (i.e., the roof)?

The tentorium cerebelli

Which three structures pass through the internal auditory meatus (i.e., the opening in the posterior aspect of the petrous part of the temporal bone)?

1. CN VII (the abducens nerve)
2. CN VIII (the vestibulocochlear nerve)
3. The labyrinthine artery

Which cranial foramen lies at the junction of the petrous part of the temporal bone and the occipital bone?

The jugular foramen

Which six structures pass through the jugular foramen?

1. CN IX (the glossopharyngeal nerve)
2. CN X (the vagus nerve)
3. CN XI (the accessory nerve)
4. The internal jugular vein (superior bulb)
5. The sigmoid sinus
6. The inferior petrosal sinus

Where is the hypoglossal canal in relation to the jugular foramen?

The hypoglossal canal lies just medial to the jugular foramen.

Which nerve passes through the hypoglossal canal?

CN XII (the hypoglossal nerve)

Which large opening lies in the posterior midline floor of the posterior fossa?

The foramen magnum

Which structures pass through the foramen magnum?

1. The medulla oblongata (i.e., the lower aspect of the brain stem)
2. CN XI (the spinal accessory nerve)
3. The vertebral arteries
4. The venous plexus of the vertebral canal
5. The anterior and posterior spinal arteries

What is the name of the bony "ramp" just anterior to the foramen magnum?

The clivus

Which small opening may be present lateral to the foramen magnum?

The condyloid foramen

Which structure passes through the condyloid foramen?

The condyloid emissary vein

Which structures pass through the mastoid foramen?

The mastoid emissary vein and a branch of the occipital artery

What is the name of the midline crest on the inside of the occipital bone?

The internal occipital crest

Which structure attaches to this crest?

The falx cerebelli (i.e., the dural fold that separates the cerebellar hemispheres)

What is the posterior termination of the internal occipital crest?

The internal occipital protuberance

Which structures are transmitted in the grooves that project laterally from the internal occipital protuberance along the occipital bone?

The transverse sinuses

FACIAL CRANIUM

Label the structures shown on the following anterior view of the skull:

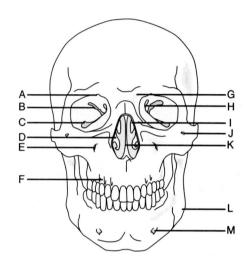

A = Supraorbital notch
B = Superior orbital fissure
C = Inferior orbital fissure
D = Inferior nasal concha
E = Infraorbital foramen
F = Alveolar process
G = Glabella
H = Optic canal
I = Middle nasal concha
J = Zygomaticofacial foramen
K = Nasal septum
L = Angle of the mandible
M = Mental foramen

What is the smooth median prominence of the frontal bone called?

The glabella

Which bone located between the orbits contains the cribriform plate and a perpendicular plate?

The ethmoid bone

**Which structures pass
through the:**

    Infraorbital foramen?

The infraorbital nerve (a continuation of
CN V_2), the infraorbital artery, and the
infraorbital vein

    **Zygomaticofacial
    foramen?**

The zygomaticofacial nerve

    Mental foramen?

The mental nerve, the mental artery, and
the mental vein

Orbit

**Which bones form the
margins of the orbit:**

    Superiorly?

The frontal bone (orbital plate)

    Laterally?

The zygomatic bone and zygomatic
process of the frontal bone

    Inferiorly?

The maxilla and zygomatic bones

    Medially?

The ethmoid, lacrimal, sphenoid, and
frontal bones

**Which two fissures form a
communication between
the intracranial space and
the orbit?**

The superior orbital fissure
(communicates with the middle cranial
fossa) and the inferior orbital fissure (communicates with the infratemporal fossa)

**Which structures pass into
the orbit from the infratemporal fossa via the inferior
orbital fissure?**

The zygomatic branch of CN V_2 (the
maxillary division of the trigeminal nerve)
and the infraorbital artery

**Which structures pass
through the supraorbital
notch?**

CN V_1 (the opthalmic division of the
trigeminal nerve, or the supraorbital
nerve) and the supraorbital vessels

Paranasal (air) sinuses

What are paranasal sinuses?

The paranasal sinuses are air spaces within
the bones of the skull that communicate
with the nasal cavity. Don't confuse the
paranasal sinuses with the venous (dural)
sinuses, which convey venous blood in the
cranium.

List the four skull bones that have paranasal sinuses.	1. Frontal bone 2. Maxilla 3. Ethmoid bone 4. Sphenoid bone
What is the function of these sinuses?	Their function is unknown, although they are thought to lighten the skull and aid in resonation of the voice.
Which sinus, because of its location, can often lead to the spread of infection into the orbit?	The ethmoid sinus
Which sinus is susceptible to the spread of infection from a tooth?	The maxillary sinus—roots of the posterior maxillary teeth often project up into this sinus.
Which sinuses may be present at birth?	The maxillary and sphenoid sinuses

Nasal cavity

Which four bones form the roof of the nasal cavity?	The nasal bone, the frontal bone, the cribriform plate of the ethmoid bone, and the body of the sphenoid bone
Which bones form the floor of the nasal cavity (and the hard palate, or the anterior portion of the roof of the mouth)?	The palatine process of the maxilla and the horizontal plate of the palatine bone
Which eight bones form the lateral wall of the nasal cavity?	1. The ethmoid bone 2. The medial pterygoid plate 3. The perpendicular plate of the palatine bone 4. The maxilla 5. The nasal bone 6. The frontal bone 7. The lacrimal bone 8. The inferior concha
Which opening forms a communication between the nasal cavity and the nasopharynx?	The choanae, a large opening at the back of the nasal cavity
Which structure divides the nasal cavity into left and right parts?	The nasal septum

Which two bones contribute to the nasal septum?	The vomer and the perpendicular plate of the ethmoid bone
What are the three bony projections from the lateral nasal wall called?	The nasal conchae—the superior and middle conchae are part of the ethmoid bone, while the inferior nasal concha is an independent bone
What are the spaces below each of the nasal conchae called?	Meatus (e.g., the superior meatus is the space between the superior and middle conchae; the middle meatus is the space between the middle and inferior conchae; and the inferior meatus is the space below the inferior concha)
The inferior meatus contains the opening to which structure?	The nasolacrimal duct
What is the name of the space above the superior concha?	The sphenoethmoid recess
What is the rounded prominence on the wall of the middle meatus?	The ethmoid bulla
What is the hiatus semilunaris?	The curved cleft below the ethmoid bulla
What is the infundibulum?	The channel at the anterior aspect of the hiatus semilunaris
Describe the drainage of each of the following paranasal sinuses:	
The anterior ethmoid sinus	The hiatus semilunaris (via the infundibulum), located in the middle nasal meatus
The middle ethmoid sinus	The ethmoid bulla, located in the middle nasal meatus
The posterior ethmoid sinus	The superior nasal meatus
The frontal sinus	The middle nasal meatus (via the frontonasal duct, which opens into the infundibulum)
The sphenoid sinus	The sphenoethmoidal recess

Maxillary sinus	The hiatus semilunaris
Which three arteries supply branches to the nasal cavity?	The maxillary and facial arteries (branches of the external carotid artery) and the ophthalmic artery (a branch of the internal carotid artery)
What are the two primary branches of the ophthalmic artery that supply the nasal cavity?	The anterior and posterior ethmoidal arteries, which supply the lateral wall and nasal septum
What are the two branches of the maxillary artery that supply the nasal cavity?	The sphenopalatine artery (which supplies the conchae, meatus, and nasal septum) and the descending palatine artery (which also supplies the nasal septum)
Which branch of the facial artery supplies the nasal cavity?	The superior labial artery

Mandible

Where does the body of the mandible meet each of the rami?	At the angle
Name the two processes on top of each of the rami.	The coronoid process (anterior) and the condyloid process (posterior)
What is the name of the notch located between the coronoid and condyloid processes?	The mandibular notch
What is the name of the opening on the medial surface of the mandible?	The mandibular foramen
What structure does the mandibular foramen lead to?	The mandibular canal
What structures lie in the mandibular canal?	The inferior alveolar nerves and vessels

Infratemporal fossa

Which structures pass though the:

Petrotympanic fissure?	The chorda tympani (a branch of the facial nerve within the temporal bone)
The stylomastoid foramen?	The facial nerve
The greater palatine foramen?	The greater palatine nerve and vessels
The lesser palatine foramen?	The lesser palatine nerve and vessels

THE SCALP AND SUPERFICIAL AND DEEP FACE

SCALP

What is the scalp?	The skin and fascia that covers the neurocranium
What are the five layers of the scalp?	**SCALP** **S**kin **C**onnective tissue **A**poneurosis (galea aponeurotica) **L**oose connective tissue **P**ericranium
Branches of which artery constitute the major blood supply of the scalp?	The external carotid artery
What are these branches of the external carotid artery that supply the scalp?	The superficial temporal, posterior auricular, and occipital arteries
Which branches of the internal carotid artery supply the scalp?	The supratrochlear and supraorbital arteries (via the ophthalmic artery)
What is unique about the veins of the scalp?	They have no valves.

What connects the veins of the scalp with the veins of the skull bones and the veins within the cranium?	Emissary veins

NOSE

What is the medical term for nostrils?	Nares
What is the name of the cartilaginous part of the external nose that surrounds each naris?	The ala nasi ("wing of the nose")
What is the dilated part of the nostril called?	The vestibule
What three effects does the nose have on inspired air?	Warming, moistening, and filtering

MUSCLES OF FACIAL EXPRESSION

Label the muscles of facial expression on the following figure:

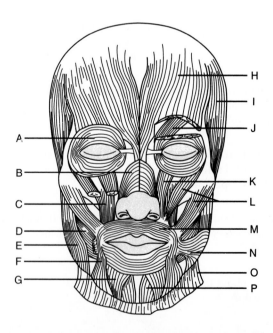

A = Orbicularis oculi
B = Nasalis
C = Levator anguli oris
D = Buccinator
E = Masseter
F = Depressor anguli oris
G = Depressor labii inferioris
H = Frontalis
I = Temporalis
J = Corrugator (supercilii)
K = Levator labii superioris
L = Zygomaticus major and minor
M = Orbicularis oris
N = Risorius
O = Platysma
P = Mentalis

What are the muscles of facial expression derived from?	The mesoderm of the second pharyngeal arch (i.e., the hyoid arch)
Dysfunction of which muscle results in difficulty with blinking?	The orbicularis oculi (closes eyes)
Which muscle is primarily responsible for raising the eyebrows?	The frontalis muscle
Which nerve innervates the muscles of facial expression?	CN VII (the facial nerve)

TEMPOROMANDIBULAR JOINT (TMJ) AND MUSCLES OF MASTICATION

Temporomandibular joint (TMJ)

What type of joint is the TMJ?	A synovial joint
Which two types of movements are provided by the TMJ?	Hinge movement and sliding movement
What are the articular surfaces of the TMJ?	The articular tubercle and mandibular fossa of the temporal bone and the condyloid process of the mandible

Name the three ligaments of the TMJ.	1. The lateral temporomandibular ligament 2. The sphenomandibular ligament 3. The stylomandibular ligament
Which ligament reinforces the TMJ by stretching from the tubercle on the zygoma to the neck of the mandible?	The lateral temporomandibular ligament
Which ligament reinforces the TMJ by stretching from the spine of the sphenoid bone to the lingula of the mandible?	The sphenomandibular ligament
Dislocation of the TMJ usually occurs in which direction?	Anteriorly

Muscles of mastication

What are the four muscles of mastication?	1. The masseter muscle 2. The temporalis muscle 3. The medial pterygoid muscle 4. The lateral pterygoid muscle
Describe the origin and insertion for each of the following:	
Masseter muscle	**Origin:** The lower border and medial surface of the zygomatic arch **Insertion:** The lateral surface of the coronoid process, ramus, and angle of the mandible
Temporalis muscle	**Origin:** The floor of the temporal fossa **Insertion:** The coronoid process and ramus of the mandible
Lateral pterygoid muscle (superior head)	**Origin:** The infratemporal surface of the sphenoid bone **Insertion:** The neck of the mandible
Lateral pterygoid muscle (inferior head)	**Origin:** The lateral surface of the lateral pterygoid plate

	Insertion: The articular disk and capsule of the TMJ
Medial pterygoid muscle	**Origin:** The tuber of the maxilla, the medial surface of the lateral pterygoid plate, and the pyramidal process of the palatine bone **Insertion:** The medial surface of the angle and the ramus of the mandible

Which muscle is responsible for:

Closing the jaw and pro-truding the mandible?	The medial pterygoid muscle
Opening the jaw and pro-truding the mandible?	The lateral pterygoid muscle

Which muscle is the primary actor in:

Depressing the mandible at the TMJ?	The lateral pterygoid muscle
Retracting the mandible?	The temporalis muscle

Which three muscles act to elevate the mandible at the TMJ?	The temporalis, masseter, and medial pterygoid muscles
Which two muscles act to close the jaw and retract the mandible?	The masseter muscle and the temporalis muscle
Which arteries supply the muscles of mastication?	Small branches from the maxillary artery (sometimes called the pterygoid branches)

PAROTID GLAND

What is the name of the largest of the three sets of salivary glands?	The parotid gland

What are the boundaries of the space containing the parotid gland:

Anteriorly?	The mandible and muscles of mastication

Posteriorly?	The external auditory meatus and mastoid process
Medially?	The styloid process
Superiorly?	The zygomatic arch
Which structure divides the parotid gland into superficial and deep parts?	CN VII (the facial nerve)
Which part of the parotid gland extends upward behind the TMJ?	The glenoid process
Which part of the parotid gland extends anteriorly, superficial to the masseter muscle?	The facial process
What is the name of the extension of the parotid gland between the medial pterygoid muscle and the mandible?	The pterygoid process
What is the name of the major duct of the parotid gland?	The parotid duct (Stensen's duct)
Where does the parotid duct begin?	At the facial process of the parotid gland
The parotid duct pierces which muscle on its course anteriorly?	The buccinator
Where does the parotid duct open into the oral cavity?	Opposite the second upper molar tooth
Which three major structures traverse the parotid gland?	1. CN VII (the facial nerve) 2. The retromandibular vein 3. The external carotid artery

From which ganglion do the parasympathetic fibers that supply the parotid gland originate?

The otic ganglion

What is the source of parasympathetic fibers to the otic ganglion?

Parasympathetic fibers originate in the inferior salivary nucleus of CN IX (the glossopharyngeal nerve), follow the tympanic branch, and then travel to the otic ganglion via the lesser petrosal nerve.

Which nerve transmits postganglionic parasympathetic fibers to the parotid gland?

After passing through the otic ganglion, the postganglionic parasympathetic fibers are transmitted to the parotid gland via the auriculotemporal nerve.

Sympathetic innervation to the parotid gland follows which structure?

The external carotid artery

SUBMANDIBULAR REGION

Name the two sets of salivary glands that lie in the submandibular region.

1. Submandibular glands
2. Sublingual glands

Of the three sets of salivary glands, which is the smallest?

The sublingual glands

Where do the ducts of the sublingual salivary glands open into the mouth?

The mucous membrane of the floor of the mouth

Which structure separates the parotid gland and the submandibular glands?

The stylomandibular ligament

Which duct drains the submandibular gland?

Wharton's duct

Wharton's duct lies between which two structures?

The sublingual glands and the genioglossus muscle

Where does Wharton's duct empty?

Alongside the frenulum of the tongue

Which major blood vessels run in the submandibular region?

The facial and lingual arteries (both branches of the external carotid artery)

What is the source of para-sympathetic innervation to the submandibular and sublingual salivary glands?

CN VII (the facial nerve), via the chorda tympani and lingual nerve

INNERVATION OF THE FACE

Which cranial nerve provides motor innervation to the face?

CN VII (the facial nerve)

Identify the five terminal branches of CN VII on the following figure:

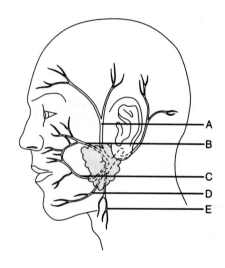

A = Temporal branch
B = Zygomatic branch
C = Buccal branch
D = Mandibular branch
E = Cervical branch

Within which structure does CN VII divide into its branches?

The parotid gland

Which cranial nerve provides sensory innervation to the face?

CN V (the trigeminal nerve)

Identify the nerves that provide sensory innervation to the face on the following figure:

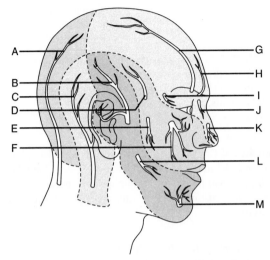

A = Greater occipital nerve
B = Auriculotemporal nerve
C = Lesser occipital nerve
D = Zygomaticotemporal nerve
E = Zygomaticofacial nerve
F = Infraorbital nerve
G = Supraorbital nerve
H = Supratrochlear nerve
I = Lacrimal nerve
J = Infratrochlear nerve
K = External nasal nerve
L = Buccal nerve
M = Mental nerve

The supraorbital and supratrochlear nerves are branches of which nerve?

The frontal nerve

What structures are innervated by the supraorbital and supratrochlear nerves?

The scalp, forehead, and upper eyelid

Which nerve provides sensory innervation to the eye and the septum, lateral walls, and tip of the nose?

The nasociliary nerve, a branch of CN V_1 (the opthalmic division of the trigeminal nerve)

Describe the branches and termination of the nasociliary nerve.	After giving off the posterior ethmoidal nerve, the ciliary nerves to the iris, and a communicating branch to the ciliary ganglion, the nasociliary nerve ends in the anterior ethmoidal nerve and the infraorbital nerve.
Which branch of the nasociliary nerve provides sensory innervation to the septum, lateral walls, and tip of the nose?	The anterior ethmoidal nerve
Which branches of CN V_2 (the maxillary division of the trigeminal nerve) provide sensory innervation to the nasal cavity?	1. Posterior inferior lateral nasal nerve 2. Posterior superior lateral nasal nerve 3. Nasopalatine nerve 4. Anterior superior alveolar nerve
Which cranial nerve provides special sensory innervation to the nose?	CN I (the olfactory nerve)—CN I passes through the opening of the cribriform plate of the ethmoid bone on the way to the olfactory bulbs

Describe the sensory innervation of the:

Roof of the mouth	The greater palatine and nasopalatine nerves (branches of CN V_2, the maxillary division of the trigeminal nerve)
Floor of the mouth	The lingual nerve (a branch of CN V_3, the mandibular division of the trigeminal nerve)
Cheek	The buccal nerve (also a branch of CN V_3)

VASCULATURE OF THE FACE

Arteries

What is the source of arterial blood to the face?	The external carotid artery

Name the eight branches of the external carotid artery, from proximal to distal.

1. Superior thyroid artery
2. Ascending pharyngeal artery
3. Lingual artery
4. Facial artery
5. Occipital artery
6. Posterior auricular artery
7. Superficial temporal artery
8. Maxillary artery

Facial artery

Describe the course of the facial artery in the sub-mandibular region.

The facial artery arises from the external carotid artery above the hyoid bone, ascends deep to the digastric and stylohyoid muscles and then behind the submandibular gland, hooks around the inferior border of the mandibular body and then enters the anterior margin of the masseter muscle.

Where can the pulse of the facial artery be easily palpated?

Just inferior to the mandible at the anterior border of the masseter muscle

Name three branches of the facial artery.

1. Inferior labial artery
2. Superior labial artery
3. Lateral nasal artery

Maxillary artery

Where does the maxillary artery branch from the external carotid artery?

At the posterior border of the ramus of the mandible, within the parotid gland

The maxillary artery is divided into three parts by which muscle?

The lateral pterygoid muscle

Where does the pterygo-palatine part of the maxillary artery run in relation to the lateral pterygoid muscle?

Between the two heads

Identify the branches of the maxillary artery on the following figure:

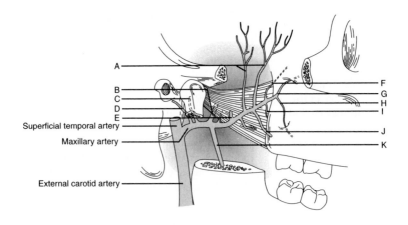

A = Deep temporal artery
B = Middle meningeal artery
C = Anterior tympanic artery
D = Deep auricular artery
E = Masseter artery
F = Sphenopalatine artery
G = Infraorbital artery
H = Posterior superior alveolar artery
I = Descending palatine artery
J = Buccal artery
K = Inferior alveolar artery

Name the five branches of the maxillary artery within the infratemporal fossa.

1. Deep auricular artery
2. Anterior tympanic artery
3. Middle meningeal artery
4. Accessory meningeal artery
5. Inferior alveolar artery

Which branch of the maxillary artery supplies the:

External auditory meatus?   The deep auricular artery

Tympanic membrane?   The anterior tympanic artery

Damage to which artery results in an epidural hematoma?

The middle meningeal artery (most often damaged in fractures of the temporal bone)

How does the middle meningeal artery enter the skull?

Through the foramen spinosum

Which branch of the maxillary artery supplies the chin and lower teeth?

The inferior alveolar artery

Name the six major branches of the maxillary artery in the pterygopalatine fossa.

1. Posterior superior alveolar artery
2. Infraorbital artery
3. Descending palatine artery
4. Artery of the pterygoid canal
5. Pharyngeal artery
6. Sphenopalatine artery

Describe the course of the sphenopalatine artery and the structures it supplies.

The sphenopalatine artery leaves the pterygopalatine fossa, passes through the pterygopalatine foramen, and enters the nasal cavity, where it supplies the conchae, meatus, and nasal septum.

Superficial temporal artery

Where can the superficial temporal arterial pulse be palpated?

Just anterior to the auricle of the external ear

Which nerve accompanies the superficial temporal artery?

The auriculotemporal nerve

The superficial temporal artery gives rise to the transverse facial artery. Between which two structures does the transverse facial artery pass?

The zygomatic arch above and the parotid duct below

Veins

Describe three pathways for venous drainage in the face and scalp.

1. Facial vein to the retromandibular vein to the external jugular vein
2. Plexuses within the face to the external jugular vein
3. Venous (dural) sinuses to the internal jugular vein

Describe the origin of the retromandibular vein.

The retromandibular vein is formed when the superficial temporal and maxillary veins unite.

Describe the origin of the facial vein.

The supraorbital and supratrochlear veins join to form the angular vein, which becomes the facial vein at the lower margin of the orbit.

The facial vein joins with which structure to form the internal jugular vein?

The retromandibular vein

What are the two major avenues of venous drainage from the infratemporal fossa?

The maxillary vein and the pterygoid venous plexus

How does the pterygoid venous plexus communicate with the facial vein?

Via the deep facial vein

Lymphatics

What are the three main lymphatic chains in the face?

The anterior, lateral, and occipital chains

Where does the anterior lymphatic chain drain?

To the submandibular and submental nodes, then to the deep cervical nodes

Where does the lateral lymphatic chain drain?

To the superficial parotid and deep parotid nodes, and then to the deep cervical nodes

Where does the occipital lymphatic chain drain?

To the occipital and retroauricular nodes, and then to the deep cervical nodes

THE ORAL CAVITY

PALATE

What forms the roof of the mouth?

The hard palate (anteriorly) and the soft palate (posteriorly)

Which bones comprise the hard palate?

The palatine processes of the maxilla and the horizontal plates of the palatine bones

Where is the incisive foramen located?

Posterior to the central incisor teeth

Which structures pass through the incisive foramen?	The greater and lesser palatine arteries (branches of the sphenopalatine artery) and the nasopalatine nerve

TEETH

What is the normal number of adult teeth?	32 (8 incisors, 4 canines, 8 premolars, and 12 molars)
Which nerves provide innervation to the maxillary teeth?	The anterior, middle, and posterior superior alveolar branches of CN V_3 (the maxillary division of the trigeminal nerve)
Which nerve innervates the mandibular teeth?	The inferior alveolar branch of the mandibular nerve
What mechanism is responsible when a patient experiences:	
Ear pain as a result of a lower jaw infection?	Irritation of the mandibular nerve
Symptoms of sinusitis as a result of a tooth infection?	Irritation of the maxillary nerve

TONGUE

What is the name of the mucous membrane fold on the midline undersurface of the tongue?	The frenulum
What is ankyloglossia?	An abnormally short frenulum (can lead to speech impediment)
What nerve provides for taste sensation on the anterior two thirds of the tongue, in addition to providing parasympathetic innervation to the submandibular and sublingual salivary glands and the lacrimal glands?	The chorda tympani, a branch of CN VII (the facial nerve)
Which nerve provides for taste sensation on the posterior third of the tongue?	CN IX (the glossopharyngeal nerve)

Which nerve provides sensory innervation to the tongue?	CN V_3 (the mandibular division of the trigeminal nerve)

Musculature of the tongue

What is the function of the intrinsic tongue muscles?	They help the tongue maintain its shape.
Which four muscles comprise the extrinsic musculature of the tongue?	1. Genioglossus 2. Hyoglossus 3. Styloglossus 4. Palatoglossus

What is the origin, insertion, and action of the:

Genioglossus muscle?	**Origin:** The genial tubercle of the mandible **Insertion:** The inferior aspect of the tongue and the body of the hyoid bone **Action:** Protrudes and depresses the tongue
Hyoglossus muscle?	**Origin:** The body of the greater horn of the hyoid bone **Insertion:** The side and inferior aspect of the tongue **Action:** Depresses and retracts the tongue
Styloglossus muscle?	**Origin:** The styloid process **Insertion:** The side and inferior aspect of the tongue **Action:** Retracts and elevates tongue
Palatoglossus muscle?	**Origin:** Aponeuroses of the soft palate **Insertion:** The dorsolateral side of the tongue **Action:** Elevates the tongue
Which one of the extrinsic muscles of the tongue is not innervated by CN XII (the hypoglossal nerve)?	The palatoglossus [this muscle is innervated by CN X (the vagus nerve) via the pharyngeal plexus]
Lesions of CN XII cause the tongue to deviate toward which side?	Toward the side of the lesion (this is known as the "wheelbarrow effect;" think of what happens when you push a wheelbarrow with one hand—to which side does it tend to deviate?)

Vasculature of the tongue

**What is the arterial supply
to the tongue?**

The lingual branch of the external carotid
artery, the ascending pharyngeal artery,
and branches of the facial artery

**What is the lymphatic
drainage from the:**

    **Anterior third (tip) of the
    tongue?**

The submental nodes

    **Posterior two thirds of
    the tongue?**

The submental nodes and the
submandibular nodes (to the deep
cervical lymphatic chain)

**What are the lymph nodules
located under the posterior
tongue called?**

The lingual tonsils

THE PHARYNX

What is the pharynx?

A muscular tube through which food and
water pass to the esophagus and air passes
to the larynx, trachea, and lungs

**What are the superior and
inferior borders of the
pharynx?**

The pharynx extends from the base of the
skull to vertebra C6

PHARYNGEAL MUSCLES

**What two groups of muscles
comprise the pharynx?**

1. External circular layer (the
   constrictors)
2. Internal longitudinal layer

**Pharyngeal constrictor
(external) muscles**

**Name the three constrictor
muscles, from interior to
exterior.**

1. Superior constrictor
2. Middle constrictor
3. Inferior constrictor

**What is the action of the
constrictors?**

By constricting in a coordinated fashion,
these muscles push food into the
esophagus. Constriction is under
autonomic control.

Overlapping of the pharyngeal constrictors creates four gaps that allow structures to access the pharynx. Where are these gaps located?

1. Superior to the superior constrictor
2. Between the superior and middle constrictors
3. Between the middle and inferior constrictors
4. Inferior to the inferior constrictor

What is the origin of the:

Superior constrictor?

The pterygoid hamulus, pterygomandibular raphe, and posterior mylohyoid line of the mandible

Middle constrictor?

The stylohyoid ligament and hyoid bone

Inferior constrictor?

The thyroid and cricoid cartilages of the larynx

What is the common insertion for the three constrictor muscles?

The median raphe of the pharynx (i.e., the midline in the posterior of the pharynx)

What nerve innervates all of the constrictor muscles?

The pharyngeal and superior laryngeal branches of CN X (the vagus nerve), via the pharyngeal plexus

Longitudinal (internal) muscles

Name the three longitudinal (internal) muscles of the pharynx.

1. Palatopharyngeus muscle
2. Salpingopharyngeus muscle
3. Stylopharyngeus muscle

What is the action of the longitudinal muscles?

These muscles raise the pharynx and larynx during swallowing and speaking.

What is the origin of the:

Palatopharyngeus muscle?

The hard palate (*palato-*)

Salpingopharyngeus muscle?

The cartilaginous eustachian tube (*salpinx* means "tube" in Latin)

Stylopharyngeus muscle?

The styloid process (*stylo-*)

What is the common insertion of the longitudinal muscles?

The posterior and superior border of the thyroid cartilage

What is the innervation of the longitudinal muscles?

The palatopharyngeus muscle and the salpingopharyngeus muscle receive innervation from CN X (the vagus nerve) via the pharyngeal plexus. The stylopharyngeus muscle is innervated by CN IX (the glossopharyngeal nerve).

PHARYNGEAL REGIONS

What are the three divisions of the pharynx?

The nasopharynx, oropharynx, and laryngopharynx

Identify the labeled structures on the following posterior view of the pharyngeal region:

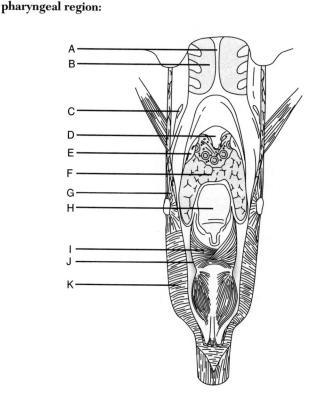

A = Nasal septum
B = Nasal cavity
C = Eustachian (auditory) tube
D = Uvula
E = Tonsil

F = Tongue
G = Middle constrictor muscle
H = Epiglottis
I = Arytenoid muscle
J = Piriform recess
K = Inferior constrictor muscle

Nasopharynx

What structure forms the boundary between the nasal cavity and the nasopharynx?

The choanae, a large opening at the posterior extent of the nasal cavity

Which lymphoid tissue collection protrudes from the posterior roof of the nasopharynx?

The pharyngeal tonsils

Which structure connects the tympanic cavity with the nasopharynx?

The eustachian tube

Where in the nasopharynx is the opening of the eustachian tube?

In the lateral wall

Which muscle attaches to the eustachian tube?

The salpingopharyngeus muscle

Oropharynx

What forms the border between the oral cavity and the oropharynx?

The palatoglossal arch (formed by underlying muscle of the same name)

What are the superior and inferior borders of the oropharynx?

The soft palate (superiorly) and the superior border of the epiglottis (inferiorly)

Which two folds bound the oropharynx laterally?

The palatoglossal and palatopharyngeal arches

What lies between these two folds?

The palatine tonsils (within the tonsillar sinus)

What is the name of the inferior projection from the midline of the soft palate?

The uvula (also known as "that little thing at the back of your throat")

What is the sensory inner-
vation of the oropharynx?

CN IX (the glossopharyngeal nerve)

List the five muscles of the
soft palate.

1. Tensor veli palatini muscle
2. Levator veli palatini muscle
3. Palatoglossus muscle
4. Palatopharyngeus muscle
5. Musculus uvulae muscle

Which muscle is responsible
for elevating and retracting
the soft palate?

The levator veli palatini

Laryngopharynx

What is the laryngopharynx?

The portion of the pharynx that lies
posterior to the larynx

Posteriorly, the laryngo-
pharynx is related to which
vertebrae?

Vertebrae C4–C6

What are the superior and
inferior margins of the
laryngopharynx?

Superior: The upper border of the
epiglottis
Inferior: The lower border of the cricoid
cartilage (i.e., the beginning of the
trachea)

THE EYE AND ADNEXA

EYEBALL

Identify the structures on
the following figure of the
eye:

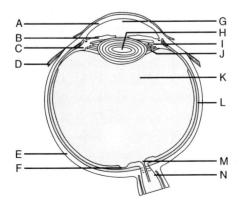

A = Cornea
B = Iris
C = Ciliary body
D = Lateral rectus muscle
E = Sclera
F = Macula (a part of the retina)
G = Aqueous humor
H = Lens
I = Ciliary processes
J = Suspensory ligament (zonular fibers)
K = Vitreous body
L = Choroid
M = Optic disc
N = CN II (the optic nerve)

Chambers of the eye

The lens lies between which two structures?

The iris (anteriorly) and the vitreous body (posteriorly)

The iris divides the space between the lens and cornea into which chambers?

The anterior and posterior chambers

What fills these chambers?

Aqueous humor

Trace the flow of aqueous humor.

Made by the ciliary processes, the aqueous humor enters the posterior chamber, flows into the anterior chamber (via the pupil), and is drained from the anterior chamber via the canal of Schlemm.

Obstruction of the canal of Schlemm can lead to which clinical condition?

Glaucoma

What are the consequences of glaucoma?

The resultant increase in intraocular pressure can cause retinal damage and blindness.

Ocular tunics

What are the three ocular tunics?

The fibrous tunic, the vascular tunic, and the retina

Fibrous tunic

What two structures form the fibrous tunic of the eyeball?

The sclera (covers the posterior five sixths of the eyeball) and the cornea (covers the anterior one sixth of the eyeball)

Which structures pierce the sclera?

1. CN II (the optic nerve)
2. The central artery of the retina, a branch of the ophthalmic artery (encased within the dural sheath of CN II)
3. The ciliary nerve, artery, and vein

What is the function of the cornea?

Refraction of light

What is the name of the site where the cornea and the sclera meet?

The limbus

Vascular tunic

Which three structures comprise the middle vascular tunic of the eyeball?

The choroid, the ciliary body (i.e., the ciliary muscle and the ciliary processes), and the iris

Which structure is responsible for nourishing the retina?

The choroid

What action changes the convexity of the lens?

Contraction of the ciliary muscle

When the lens is focusing on distant objects, which change occurs within it?

It flattens. Flattening of the lens is achieved by relaxation of the ciliary muscle, which leads to contraction of the suspensory ligament.

What is the nerve supply to the ciliary muscle?

Parasympathetic fibers from CN III (the oculomotor nerve), via the short ciliary nerves (which in turn, are sent from the ciliary ganglion)

What is the name of the central pigmented diaphragm in the middle eye layer?

The iris (i.e., the colored part of the eye)

What is the central aperture of the iris called?

The pupil

Which type of fibers innervate:

The sphincter pupillae (i.e., the circular muscle fibers of the iris)?

Parasympathetic fibers from CN III (the oculomotor nerve), via the short ciliary nerves from the ciliary ganglion

| The dilator pupillae (i.e., the radial muscle fibers of the iris)? | Sympathetic fibers, via the long ciliary nerve |

Retina

| What are the two layers of the retina? | The pigmented retina and the neural retina |

| Where are the photo-receptors (i.e., rods and cones) found? | In the neural retina |

| Which structures are specialized for vision in dim light? | Rods |

| Which structures are specialized for visual acuity and color vision? | Cones |

| The greatest visual acuity is found on which portion of the retina? | The macula (near the center) |

| What is the name of the central depression in the macula? | The fovea centralis (contains only cones) |

| Axons from ganglion cells of the retina converge to form which structure? | CN II (the optic nerve) |

| What is the name of the origin of CN II on the retina? | The optic disc |

| What is the center of the optic disc called? | The optic cup |

Innervation of the eye

| What is the name of the parasympathetic ganglion in the posterior orbit, lateral to the optic nerve? | The ciliary ganglion |

The ciliary ganglion transmits parasympathetic fibers to which structures via the short ciliary nerves?

The sphincter pupillae and ciliary muscles

Which nerve provides the sense of sight?

CN II (the optic nerve)

Why is CN II unusual?

It is invested by all three layers of meninges (i.e., the pia mater, the arachnoid mater, and the dura mater) throughout its course. Therefore, increased intracranial pressure (as may occur with tumors or bleeding) is transmitted to the back of the orbit via the subarachnoid space, which contains cerebrospinal fluid (CSF). This causes characteristic retinal abnormalities that may be noted on ophthalmoscopic examination.

Where does the optic nerve from one eye join the optic nerve from the opposite eye?

At the optic chiasm

Vasculature of the eye

The optic disc is pierced by which blood vessel?

The central artery of the retina

What can result from occlusion of the central artery of the retina?

Instant blindness

Which feature of the central artery of the retina makes its occlusion such an emergency?

It is an end artery [i.e., the area supplied by the central artery of the retina has no collateral (alternative) circulation].

What is the venous drainage from the orbit?

The superior and inferior ophthalmic veins, draining into the cavernous sinus [i.e., one of the venous (dural) sinuses]

Musculature of the eye

**Define the action of each
of the following muscles:**

Superior rectus muscle	**In isolation:** Moves the eyeball superiorly and laterally **In conjunction with the inferior oblique muscle:** Moves the eyeball superiorly
Inferior rectus muscle	**In isolation:** Moves the eyeball inferiorly and medially **In conjunction with the superior oblique muscle:** Moves the eyeball inferiorly
Medial rectus muscle	Adducts the eyeball
Lateral rectus muscle	Abducts the eyeball
Superior oblique muscle	**In isolation:** Moves the eyeball inferiorly and laterally **In conjunction with the inferior rectus muscle:** Moves the eyeball inferiorly
Inferior oblique muscle	**In isolation:** Moves the eyeball superiorly and medially **In conjunction with the superior rectus muscle:** Moves the eyeball superiorly
How do the oblique muscles move the eyeball?	Medially as well as in the vertical plane
How do they move the eye straight up and down?	To move the eye straight up or down, the obliques recruit the recti muscles—the inferior oblique works with the superior rectus, and the superior oblique works with the inferior rectus.
Turning the eye medially and looking up and down tests which muscles?	The superior and inferior obliques
Turning the eye laterally and looking up and down tests which muscles?	The superior and inferior recti

What is the origin, course, and insertion of the superior oblique muscle?

From the body of the sphenoid bone, the muscle forms a tendon that runs anteriorly to reach the trochlea, where it turns posteriorly and courses laterally to insert on the sclera beneath the superior rectus.

What is the origin and insertion of the inferior oblique muscle?

Origin: The floor of the orbit
Insertion: The sclera, beneath the lateral rectus

EYELIDS

What is the name of the longitudinal opening between the upper and lower eyelids?

The palpebral fissure

What is the name of the mucous membrane lining the inner eyelids and anterior surface of the eyeball?

The conjunctiva

Which muscle closes the eyelid?

The orbicularis oculi

Which muscle opens the eyelid?

The levator palpebrae superioris (assisted by the superior tarsal muscle)

What is the origin and insertion of the levator palpebrae superioris muscle?

Origin: The lesser wing of the sphenoid bone
Insertion: The skin of the upper eyelid

Which nerve innervates the levator palpebrae superioris muscle?

The superior division of CN III (the oculomotor nerve)

LACRIMAL APPARATUS

Tears are produced by which gland?

The lacrimal gland

Trace the flow of tears from the lacrimal gland to the nose.

Tears from the lacrimal gland enter the eye at the upper orbit, circulate across the cornea, gather in the lacus lacrimalis, and enter the lacrimal canaliculi, which open into the lacrimal sac. The nasolacrimal duct then empties into the inferior meatus of the nose.

What is the name of the opening into the lacrimal canaliculi, located in the medial eye?	The punctum lacrimalis
Which structure passes through the nasolacrimal canal?	The nasolacrimal duct
What is the innervation of the lacrimal gland?	Parasympathetic fibers from CN VII (the facial nerve) travel from the lacrimal nucleus to the pterygopalatine ganglion, and then to CN V_2 (the maxillary division of the trigeminal nerve). From CN V_2, the fibers travel to the lacrimal gland via the lacrimal nerve. Note: Fibers travel with CN V_2 but originate with CN VII.
In addition to the lacrimal gland, what other structures are innervated by the lacrimal nerve?	The conjunctiva and the skin of the upper eyelid

THE EAR

What are the three general parts of the ear?	The external ear, the middle ear (tympanic cavity), and the inner ear (labyrinth)
The sensory organs for hearing and balance lie within which part?	The inner ear

EXTERNAL EAR

Which two structures make up the external ear?	The auricle and external auditory meatus
What is the function of the auricle?	It collects sound vibrations.

Identify the labeled structures on the following figure of the auricle:

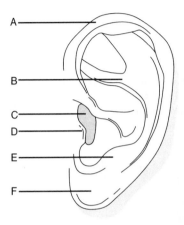

A = Helix
B = Antihelix
C = Opening of the external auditory
      meatus
D = Tragus
E = Antitragus
F = Lobule

Which nerves innervate the auricle?

1. The auricular nerve, a branch of CN X (the vagus nerve)
2. The greater auricular and lesser occipital nerves (branches from the cervical plexus)
3. The auriculotemporal nerve, a branch of CN V_3 (the mandibular branch of the trigeminal nerve)

Which portion of the external auditory meatus is formed from cartilage?

The external third

Which nerves innervate the external auditory meatus?

The auriculotemporal and auricular nerves

Which three arteries supply the auricle and external auditory meatus?

1. The posterior auricular artery (a branch of the external carotid artery)
2. The deep auricular artery (a branch of the maxillary artery)
3. The auricular branch of the superficial temporal artery

What are the two means of venous drainage from the auricle and external auditory meatus?

The external jugular and maxillary veins, and the pterygoid plexus

Which structure lies at the end of the external auditory meatus, marking the medial boundary of the external ear?

The tympanic membrane

How many layers make up the tympanic membrane?

Three (two epithelial layers and an intermediate fibrous layer)

Label the structures on the following figure of the tympanic membrane:

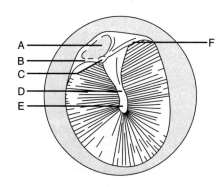

A = Flaccid part
B = Anterior mallear fold
C = Lateral process of the malleus
D = Handle of the malleus
E = Umbo
F = Posterior mallear fold

Which nerves innervate the:

Outer surface of the tympanic membrane?

The auriculotemporal and auricular nerves

Inner (medial) surface of the tympanic membrane?	The tympanic branch of CN IX (the glossopharyngeal nerve)

MIDDLE EAR (TYMPANIC CAVITY)

The middle ear is a space within which bone?	The petrous part of the temporal bone
What are the boundaries of the middle ear:	
Superiorly (i.e., the roof)?	The tegmen tympani (a thin plate of bone that is part of the petrous temporal bone)
Inferiorly (i.e., the floor)?	The jugular fossa and a thin plate of bone
Anteriorly?	The carotid canal
Posteriorly?	Mastoid air cells and the mastoid antrum
Laterally?	The tympanic membrane
Medially?	The lateral wall of the inner ear
An opening in the anterior wall of the middle ear exists for which structure?	The eustachian tube
The eustachian tube connects the middle ear with which space?	The nasopharynx
Contraction of which two muscles opens the eustachian tube?	The tensor veli palatini and salpingopharyngeus muscles
Which three nerves innervate the middle ear?	1. The auriculotemporal nerve 2. The tympanic branch of CN IX (the glossopharyngeal nerve) 3. The auricular nerve
Which three bones lie in the middle ear?	The malleus, incus, and stapes (the three ossicles, sometimes referred to as "the hammer, anvil, and stirrup," respectively, because of their shapes)
Which ossicle attaches to the tympanic membrane?	The malleus
Name the two openings on the lateral wall of the inner ear.	The oval window and the round window

Upon which opening does the foot plate of the stapes rest?	The oval window
Name the two muscles of the inner ear.	The stapedius and tensor tympani muscles
What is the origin and insertion of the:	
Stapedius muscle?	**Origin:** The posterior wall of the tympanum **Insertion:** The neck of the stapes
Tensor tympani muscle?	**Origin:** The cartilaginous portion of the eustachian tube **Insertion:** The handle of the malleus
What is the smallest skeletal muscle in the human body?	The stapedius muscle
Contraction of which muscle prevents loud noises from injuring the inner ear?	The stapedius muscle
What is the function of the tensor tympani muscle?	It dampens vibrations of the malleus.
Which nerve innervates the stapedius muscle?	The chorda tympani, a branch of CN VII (the facial nerve)
Which nerve innervates the tensor tympani muscle?	CN V_3 (the maxillary division of the trigeminal nerve)
Which three arteries supply the middle ear?	1. The posterior auricular artery 2. The anterior tympanic artery 3. The caroticotympanic artery

INNER EAR (LABYRINTH)

Which bone houses the inner ear?	The petrous portion of the temporal bone

Which three structures make up the bony labyrinth?	1. The vestibule 2. The cochlea 3. The semicircular canals
Which structures comprise the membranous labyrinth, a series of communicating sacs and ducts that is suspended in the bony labyrinth?	1. The utricle and saccule (housed in the vestibule) 2. The cochlear duct 3. The semicircular ducts
Which type of fluid is found within the bony labyrinth, surrounding the membranous labyrinth?	Perilymph
Which type of fluid fills the membranous labyrinth?	Endolymph
What is the arterial supply to the inner ear?	The labyrinthine (internal acoustic) artery, a branch of the basilar artery

Identify the labeled structures on the following illustration of the middle and inner ear:

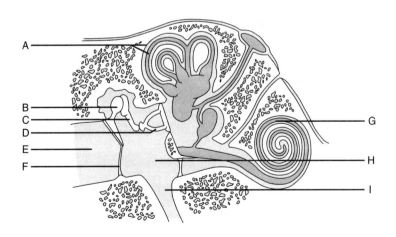

A = Semicircular canal and duct
B = Malleus
C = Incus
D = Stapes
E = External auditory meatus

F = Tympanic membrane
G = Cochlea and cochlear duct
H = Tympanic cavity
I = Eustachian tube

Cochlear apparatus

After vibrations from the external ear cause the tympanic membrane to vibrate, how are impulses conveyed to the cochlea (i.e., the organ of hearing)?

Excitation of the tympanic membrane causes the ossicles to move. The stapes (i.e., the final ossicle) then transmits vibrations to the scala vestibuli via the oval window.

Name the two compartments of the cochlea.

The scala vestibuli (above) and the scala tympani (below)

Which structure divides the cochlea into the scala vestibuli and the scala tympani?

The spiral lamina

The cochlear duct, located between the scala vestibuli and the scala tympani, contains which sensory organ?

The spiral organ of Corti

Where in the cochlea is the organ of Corti stimulated by:

Low-frequency sound waves?

Near the apex

High-frequency sound waves?

Near the base

The scala vestibuli and the scala tympani communicate at the tip of the cochlea through which structure?

The helicotrema

Identify the labeled structures on the following figure of the cochlear apparatus:

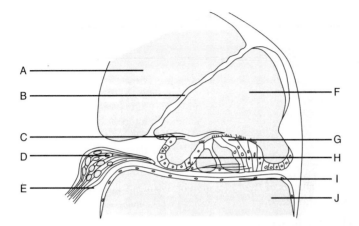

A = Scala vestibuli
B = Vestibular membrane
C = Tectorial membrane
D = Cochlear ganglion
E = Cochlear nerve
F = Cochlear duct
G = Outer hair cells
H = Inner hair cells
I = Basilar membrane
J = Scala tympani

Semicircular canals and ducts

What is the function of the semicircular canals of the inner ear?

They sense the angular acceleration of the head.

Name the three semicircular canals.

Anterior (superior), posterior, and lateral

Which sensory organs are within these canals?

Ampullae

Vestibule

What is the function of the utricle and the saccule?

They are organs that detect linear movement.

Which sensory organs are found within the utricle and saccule?

Maculae

POWER REVIEW

SKULL

Name the four major sutures of the skull.

1. Coronal suture
2. Sagittal suture
3. Squamous suture
4. Lambdoid suture

The lambdoid and sagittal sutures intersect at the lambda; the sagittal and coronal sutures intersect at the bregma.

What are the shapes of the two major fontanelles?

Anterior: Diamond-shaped
Posterior: Triangular

For each of the following cranial openings, name the bone where it is located, the structures it contains, and the structures it connects:

Optic canal

Location: Lesser wing of the sphenoid bone
Structures contained: CN II (the optic nerve), the ophthalmic artery, and the central retinal vein
Structures connected: The orbit and the middle cranial fossa

Superior orbital fissure

Location: Between the greater and lesser wings of the sphenoid bone
Structures contained: CN III (the oculomotor nerve), CN IV (the trochlear nerve), CN V_1 (the ophthalmic division of the trigeminal nerve), CN VI (the abducens nerve), and the superior ophthalmic vein
Structures connected: The middle cranial fossa and the orbit

Foramen rotundum

Location: Greater wing of the sphenoid bone
Structures contained: CN V_2 (the maxillary division of the trigeminal nerve)
Structures connected: The middle cranial fossa and the pterygopalatine fossa

Foramen ovale	**Location:** Greater wing of the sphenoid bone **Structures contained:** CN V$_3$ (the mandibular division of the trigeminal nerve) and the accessory meningeal artery **Structures connected:** The middle cranial fossa and the infratemporal fossa
Foramen spinosum	**Location:** Greater wing of the sphenoid bone **Structures contained:** The middle meningeal artery **Structures connected:** The middle cranial fossa and the infratemporal fossa
Foramen lacerum	**Location:** Between the sphenoid and the petrous part of the temporal bone **Structures contained:** Internal carotid artery, greater petrosal nerve, and deep petrosal nerve **Structures connected:** The middle and posterior cranial fossae and the neck
Which structures pass through the carotid canal?	The internal carotid artery and sympathetic nerves
Which structures pass through the foramen magnum?	The medulla oblongata, CN XI (the spinal accessory nerve), the vertebral arteries, the venous plexus, and the anterior and posterior spinal arteries
Which structures pass through the internal auditory meatus?	CN VII (the facial nerve) and CN VIII (the vestibulocochlear nerve)
Which nerve passes through the cribriform plate of the ethmoid bone?	CN I (the olfactory nerve)
What are the three bony projections from the lateral nasal wall called?	The nasal conchae

The nasal cavity is divided into four passages by the nasal conchae. Name these passages and describe their location.

The three meatus (superior, middle, and inferior) lie below the respective concha; the sphenoethmoid recess lies above the superior concha.

The inferior meatus contains the opening to which structure?

The nasolacrimal duct

Which skull bones have sinuses?

The frontal, maxillary, ethmoid, and sphenoid bones

Describe where each paranasal sinus opens into the nasal cavity:

   Anterior ethmoid sinus

Middle nasal meatus

   Middle ethmoid sinus

Middle nasal meatus

   Posterior ethmoid sinus

Superior nasal meatus

   Frontal sinus

Middle nasal meatus

   Maxillary sinus

Middle nasal meatus

   Sphenoid sinus

Sphenoethmoidal recess

SCALP AND SUPERFICIAL AND DEEP FACE

What are the five layers of the scalp?

SCALP

Skin
Connective tissue
Aponeurosis
Loose connective tissue
Periosteum

What nerve innervates the muscles of facial expression?

CN VII (the facial nerve)

Name the five terminal branches of the facial nerve.

Temporal, zygomatic, buccal, mandibular, and cervical

Name the four muscles of mastication.

1. Masseter
2. Temporalis
3. Medial pterygoid
4. Lateral pterygoids

Which nerve innervates the muscles of mastication?

CN V$_3$ (the mandibular division of the trigeminal nerve)

Sensory innervation to the skin of the face is provided by which cranial nerve?

CN V (the trigeminal nerve)

Which nerves provide for sensory innervation to the nose?

Smell: CN I (the olfactory nerve)
General sensation: CN V$_1$ and CN V$_2$ (the ophthalmic and maxillary divisions of the trigeminal nerve)

Name the eight branches of the external carotid artery, from proximal to distal.

1. Superior thyroid artery
2. Ascending pharyngeal artery
3. Lingual artery
4. Facial artery
5. Occipital artery
6. Posterior auricular artery
7. Maxillary artery
8. Superficial temporal artery

ORAL CAVITY

Which four muscles comprise the extrinsic musculature of the tongue?

1. Genioglossus
2. Hyoglossus
3. Styloglossus
4. Palatoglossus

Which tongue muscle is NOT innervated by CN XII (the hypoglossal nerve)?

The palatoglossus muscle (innervated by CN X, the vagus nerve)

Lesions of CN XII cause the tongue to deviate to which side?

The side on which the lesion is located

Describe the sensory innervation of the:

Anterior tongue

Sensation: The lingual nerve (a branch of CN V$_3$, the mandibular division of the trigeminal nerve)
Taste: The chorda tympani, a branch of CN VII (the facial nerve)

Posterior tongue	CN IX (the glossopharyngeal nerve) and CN X (the vagus nerve), for both sensation and taste

PHARYNX

What are the two major groups of pharyngeal muscles?	1. The pharyngeal constrictor (external) muscles, consisting of the superior constrictor, the middle constrictor, and the inferior constrictor 2. The longitudinal (internal) muscles, consisting of the palatopharyngeus, salpingopharyngeus, and stylopharyngeus muscles
What are the muscles of the soft palate?	1. Tensor veli palatini 2. Levator veli palatini 3. Palatoglossus 4. Palatopharyngeus 5. Musculus uvulae
Which of these muscles is not innervated by CN X (the vagus nerve)?	The tensor veli palatini (innervated by CN V_3, the maxillary division of the trigeminal nerve)
Which nerve provides sensory innervation to the:	
Nasopharynx?	CN V_2 (the maxillary division of the trigeminal nerve)
Oropharynx?	CN IX (the glossopharyngeal nerve)
Laryngopharynx?	CN X (the vagus nerve)
The palatine tonsils lie between which two arches (folds)?	The palatine tonsils are located between the palatoglossal and palatopharyngeal arches, which form the boundary of the oropharynx. Don't confuse the uvula with the palatine tonsils; the uvula hangs from the midline palate!
Which structure connects the tympanic cavity with the lateral wall of the nasopharynx?	The eustachian tube

EYE

CN III (the oculomotor nerve) innervates which muscles?	**Superior division:** The superior rectus and levator palpebrae superioris muscles **Inferior division:** The medial and inferior recti and the inferior oblique muscles; in addition, the inferior division of CN III provides parasympathetic fibers to the ciliary muscle and sphincter pupillae
CN IV (the trochlear nerve) innervates which muscles?	The superior and inferior oblique muscles
CN VI (the abducens nerve) innervates which muscle?	The lateral rectus muscle
Describe the innervation of the:	
Sphincter pupillae	Parasympathetic fibers (via CN III, the ciliary ganglion, and the short ciliary nerves)
Dilator pupillae	Sympathetic fibers via the long ciliary nerve
Which structures are specialized for:	
Vision in dim light?	Rods
Color vision?	Cones

EAR

Which three bones are contained in the middle ear?	1. The malleus (attaches to the tympanic membrane, the link to the external ear) 2. The incus 3. The stapes (attaches to the oval window of the cochlea, the link to the inner ear) Remember, the round window is the end of the cochlea.

Name the two muscles of the inner ear and the nerves that innervate them.	1. The stapedius muscle (innervated by the chorda tympani, a branch of CN VII) 2. The tensor tympani (innervated by CN V_3)
Which three structures make up the bony labyrinth of the inner ear?	The vestibule, the semicircular canals, and the cochlea
Which fluid fills the bony labyrinth?	Perilymph
Which sensory organs are found within the utricle and saccule?	Maculae (sense linear movements of the head)
What is the arterial supply to the inner ear?	The labyrinthine artery (a branch of the basilar artery)

The Central
Nervous System

INTRODUCTION

What are the three components of a neuron?	1. Cell body 2. Dendrites (carry impulses to the cell body) 3. Axon (carries impulses away from the cell body)
What is the difference between a ganglion and a nucleus?	**Ganglion:** A collection of cell bodies outside of the CNS **Nucleus:** A collection of cell bodies within the CNS
What is the difference between grey matter and white matter?	Grey matter (cell bodies) is unmyelinated; white matter (neuronal processes) is myelinated.
Which structures comprise the CNS?	The brain and the spinal cord
Which bony structures protect the CNS from injury?	The skull and the vertebral column
List three basic functions of the CNS.	1. Processing of information 2. Coordination of voluntary and involuntary behaviors 3. Higher-order cognition
How many spinal nerves and cranial nerves leave the CNS?	31 spinal nerves and 12 cranial nerves (Think, "Eat 31 flavors in 12 months")

THE BRAIN

GROSS REGIONS OF THE BRAIN

Grossly, what are the three recognizable parts of the brain?	1. Cerebrum 2. Cerebellum 3. Brain stem

Cerebrum

What is the cerebrum?

The largest part of the brain, concerned mostly with higher-order thinking

Where in the cranium is the cerebrum located?

The anterior and middle cranial fossae

How is the cerebrum divided?

The cerebrum is divided into right and left **hemispheres.** Each hemisphere contains four **lobes.** Each lobe is divided into a number of **gyri** (folds). Each gyrus contains organized collections of **neurons.**

What separates the cerebral hemispheres from one another?

The longitudinal (interhemispheric) fissure

What are gyri and sulci?

Gyri are elevated folds on the surface of the cerebral hemispheres. Sulci are the grooves that separate the gyri from each other. This architecture increases the surface area of the brain.

Identify the four lobes of the brain on the following figure:

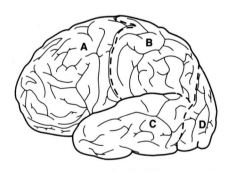

A = Frontal lobe
B = Parietal lobe
C = Temporal lobe
D = Occipital lobe

What separates the:

Frontal lobe from the parietal lobe?

The central sulcus

Parietal lobe from the occipital lobe?

The parieto-occipital sulcus

Temporal lobe from the frontal and parietal lobes?

The lateral sulcus (i.e., the sylvian fissure)

On the cerebral cortex, which region is concerned with:

Sensory information?

The **postcentral gyrus** (i.e., the part of the parietal lobe immediately adjacent to the central sulcus)

Motor signals?

The **precentral gyrus** (i.e., the part of the frontal lobe immediately adjacent to the central sulcus)

Where is the superior temporal gyrus located, and what is its role?

The superior temporal gyrus is located immediately below the Sylvian fissure and is concerned with auditory stimuli.

Where is the visual cortex located?

In the occipital lobe

Which temporal lobe structure is involved with memory?

The hippocampus

What is the cerebrum composed of?

Gray matter (superficial) and white matter

What white matter structure connects the hemispheres?

The corpus callosum

What are the basal ganglia?

Deep grey matter—specific nuclei positioned deep in the base of the cerebral hemispheres that are involved in motor function

Which five nuclei comprise the basal ganglia?

1. The caudate nucleus
2. The putamen
3. The globus pallidus

4. The substantia nigra
5. The subthalamic nucleus

Which three main arteries supply blood to the cerebrum?

1. Anterior cerebral artery
2. Middle cerebral artery
3. Posterior cerebral artery

Cerebellum

Where in the cranium is the cerebellum located?

In the posterior cranial fossa

What is the main role of the cerebellum?

Coordination of movement and postural adjustment

What clinical manifestations are characteristic in patients with cerebellar lesions?

Unsteady gait, poor coordination of movement, tremor of the hands

Which three arteries supply the cerebellum?

1. The posterior inferior cerebellar artery (a branch of the vertebral artery)
2. The anterior inferior cerebellar artery (a branch of the basilar artery)
3. The superior cerebellar artery (a branch of the basilar artery)

Brain stem

Which major structures are part of the diencephalon?

The thalamus and hypothalamus

Which structure is considered the "relay station" of the brain?

The thalamus

Which physiologic system is the hypothalamus involved with?

The endocrine system—the hypothalamus is involved in the release of hormones to the pituitary gland

What is the myelencephalon (i.e., the medulla oblongata)?

The portion of the brain stem that connects the pons with the spinal cord

What are the "pyramids?"

The two swellings of the medulla oblongata where the descending fibers from the precentral gyrus cross to the contralateral side

MENINGES OF THE BRAIN

What are meninges?

Protective coverings ("membranes") over the surface of the brain

What are the three meningeal layers, from superficial to deep?

1. Dura mater
2. Arachnoid mater
3. Pia mater

What separates the dura mater from the arachnoid mater?

The subdural space (a potential space)

What causes a subdural hematoma?

Rupture of the cerebral veins as they pass from the brain surface into the venous (dural) sinuses

What is the space between the arachnoid mater and the pia mater referred to as?

The subarachnoid space

How is the subarachnoid space different from the subdural space?

Unlike the subdural space, the subarachnoid space is a true space. It contains cerebrospinal fluid (CSF), which cushions the brain.

Dura mater

Describe the two layers of dura mater.

The outer dural layer is adherent to the bone. The inner dural layer is the true dura mater (i.e., the meningeal component).

What artery provides the major blood supply to the dura mater, and what is it a branch of?

The middle meningeal artery, a branch of the maxillary artery

Describe the course of the middle meningeal artery.

It branches from the maxillary artery in the infratemporal fossa, enters the cranium via the foramen spinosum, then runs in a groove on the inner aspect of the temporal bone.

What clinical entity results from rupture of the middle meningeal artery?

Epidural hematoma

What are the four dural folds?	1. The falx cerebri 2. The tentorium cerebelli 3. The falx cerebelli 4. The diaphragma sellae
What is the function of the dural folds?	They separate the cranial cavity into compartments.

Falx cerebri

Where is the falx cerebri located?	In the longitudinal fissure (i.e., the fissure that separates the cerebral hemispheres from one another)
What structures does the falx cerebri attach to anteriorly and posteriorly?	The crista galli of the ethmoid bone (anteriorly) and the internal occipital protuberance (posteriorly)
Which structures are found in the inferior and superior margins of the falx cerebri?	The inferior and superior sagittal sinuses, respectively

Tentorium cerebelli

Where is the tentorium cerebelli located?	The tentorium cerebelli separates the cerebrum from the brain stem and cerebellum; it lies horizontally between the occipital lobes of the cerebral hemispheres and the cerebellum. Structures below the membrane are referred to as "infratentorial."
What are the attachment points for the tentorium cerebelli:	
Anteriorly?	The clinoid processes
Laterally?	The temporal and parietal bones
Posteriorly?	The occipital bone
Which dural structure is continuous with the tentorium cerebelli?	The falx cerebri

Falx cerebelli

Where is the falx cerebelli located?	In the longitudinal fissure between the two cerebellar hemispheres (*hint:* the two falx structures run between hemispheres)

Diaphragma sellae

What structure does the diaphragma sellae form a "roof" over?

The pituitary gland

A small opening in the center of the diaphragma sella transmits which structure?

The pituitary stalk (infundibulum)

Where does the diaphragma sella attach?

To the clinoid processes

What important structure lies on top of the diaphragma sellae?

The optic chiasm

ARACHNOID MATER

What does the term "arachnoid" translate to?

"Spider-like;" the arachnoid layer is a delicate, cobweb-like structure

Pia mater

What is contained in the pia mater?

Blood vessels, blood vessels, and more blood vessels

VASCULATURE OF THE BRAIN

Arteries

Which two sets of arteries constitute the blood supply of the brain?

The internal carotid arteries and the vertebral arteries

Internal carotid arteries

Describe the course of the internal carotid artery.

The internal carotid artery enters the skull through the carotid canal (located in the petrous portion of the temporal bone), passes through the cavernous sinus, and enters the subarachnoid space.

Name the four parts of the internal carotid artery.

1. Cervical part
2. Petrous part
3. Cavernous part
4. Cerebral part

Does the internal carotid artery give off any branches in the neck before entering the cranium?

No.

What are the five major intracranial branches of the internal carotid artery?

1. Superior hypophyseal artery
2. Inferior hypophyseal artery
3. Ophthalmic artery
4. Posterior communicating artery
5. Anterior choroid artery

Which structure is the most perfused tissue in the body (per gram of tissue)?

The pituitary gland, which is supplied by the superior and inferior hypophyseal arteries, is more perfused than the heart, brain, or kidney (per gram of tissue)!

Which branch of the internal carotid artery joins the posterior cerebral artery?

The posterior communicating artery

Each internal carotid artery ends by dividing into which two arteries?

The anterior cerebral artery and the middle cerebral artery

Which is larger—the middle cerebral artery or the anterior cerebral artery?

The middle cerebral artery

Describe the course of the anterior cerebral artery.

It proceeds to the longitudinal (interhemispheric) fissure, following a curvy path around the corpus callosum.

What areas are supplied by the anterior cerebral artery?

The medial surfaces of the frontal and parietal lobes, including the area of the motor cortex that controls leg movement

Vertebral arteries

Describe the course of the vertebral arteries.

They arise from the first part of the subclavian artery, ascend through the transverse foramina of vertebrae C1–C6, and then access the intracranial space via the foramen magnum.

What structures are supplied by the vertebral arteries?

The neck muscles and the medulla oblongata

What are the clinical characteristics of a vertebral artery infarction?

Lateral medullary syndrome of Wallenberg is the result of an infarction of the lateral part of the medulla oblongata; its many features are caused by damage to the spinothalamic tract and cranial nerve origins:
1. Contralateral loss of pain and temperature sensation in the trunk and extremities
2. Ipsilateral loss of pain and temperature sensation in the face
3. Dysphagia and dysarthria
4. Vertigo, nausea, and nystagmus
5. Ipsilateral Horner's syndrome

What are the three major branches of the vertebral artery?

1. Anterior spinal artery
2. Posterior spinal artery
3. Posterior inferior cerebellar artery

Which of these branches is the largest?

The posterior inferior cerebellar artery

The vertebral arteries join on the surface of the pons to form which structure?

The basilar artery

What structures are supplied by the basilar artery?

Part of the cerebellum and the pons

What are the clinical characteristics of a basilar artery infarction?

Signs and symptoms of a basilar artery infarction are related to injury to the cerebellum and brain stem, notably the reticular activating system (which controls consciousness) and the corticospinal tracts (which are involved with motor control):
1. Coma or death
2. Quadriplegia
3. Impaired vision (as a result of visual cortex damage)
4. Disordered eye movement (as a result of cranial nerve damage)
5. Cerebellar ataxia

What are the four major branches of the basilar artery?

1. Pontine artery
2. Anterior inferior cerebellar artery
3. Labyrinthine artery
4. Superior cerebellar artery

The basilar artery ends by dividing into which vessels?

The left and right posterior cerebral arteries

What is the most common site for a berry aneurysm?

The bifurcation of the basilar artery

What is a berry aneurysm?

A congenital dilatation of a blood vessel within the brain

What is the clinical significance of a berry aneurysm?

It may rupture, leading to subarachnoid hemorrhage, or it may cause symptoms by enlarging and impinging on other structures.

Circle of Willis

What is the circle of Willis?

The circle of Willis, an arterial anastomosis network on the base of the brain, is formed by the communication of the right and left internal carotid arteries and the right and left vertebral arteries within the cranium.

Identify the arteries on the following drawing of the circle of Willis:

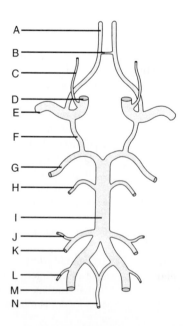

A = Anterior cerebral artery
B = Anterior communicating artery
C = Ophthalmic artery
D = Internal carotid artery
E = Middle cerebral artery
F = Posterior communicating artery
G = Posterior cerebral artery
H = Superior cerebellar artery
I = Basilar artery
J = Labyrinthine artery
K = Anterior inferior cerebellar artery
L = Posterior inferior cerebellar artery
M = Vertebral artery
N = Anterior spinal artery

Cerebral veins and venous (dural) sinuses

What is special about cerebral veins?

They are valveless.

What is the origin of the great cerebral vein (of Galen)?

It is formed by the union of the two internal cerebral veins.

What are venous sinuses?

Channels within the dura mater that route venous blood and CSF from the brain to the systemic venous circulation

Which veins, connecting the scalp and the venous sinuses, form an anastomosis within the diploë?

The diploic veins

Which veins directly connect the venous sinuses with the scalp?

The emissary veins

What vein do the venous sinuses ultimately drain into?

The internal jugular vein

Label the venous sinuses on the following figure:

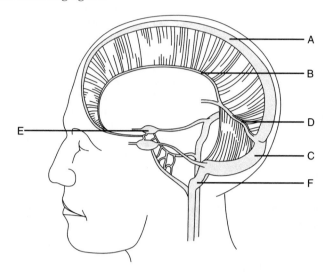

A = Cavernous sinus
B = Superior sagittal sinus
C = Inferior sagittal sinus
D = Straight sinus
E = Transverse sinus
F = Sigmoid sinus

Which sinus lies within the convex (superior) border of the falx cerebri?	The superior sagittal sinus
Which veins empty into the superior sagittal sinus?	1. The diploic veins 2. The meningeal veins 3. The emissary veins
The superior sagittal sinus ends by becoming continuous with which other sinus?	The transverse sinus
Where does this occur?	At the torcular Herophili (also known as the "confluence of sinuses"), the point where five of the six sinuses meet in the occipital area
Which sinus, located in the free (inferior) edge of the falx cerebri, is joined by the great cerebral vein (of Galen)?	The inferior sagittal sinus

What is the course of blood from the inferior sagittal sinus to the inferior jugular vein?

Inferior sagittal sinus, straight sinus, transverse sinus, sigmoid sinus, internal jugular vein

Which sinus is formed by the junction of the inferior sagittal sinus and the great cerebral vein (of Galen)?

The straight sinus

In between which dural folds does the straight sinus run?

In the line of attachment between the falx cerebri and the tentorium cerebelli

Into which structure does the straight sinus empty?

The transverse sinus

Where is the transverse sinus located?

The transverse sinus occupies the attached margin of the tentorium cerebelli.

After leaving the tentorium cerebelli, what does the transverse sinus become?

The sigmoid sinus

Describe the course of the sigmoid sinus.

The sigmoid sinus runs through a groove in the mastoid part of the temporal bone; it then runs anteriorly and inferiorly to reach the jugular foramen, where it becomes continuous with the internal jugular vein.

Which sinus is found on either side of the sella turcica?

The cavernous sinus

Which structures provide communication between the two sides of the cavernous sinus?

Intercavernous sinuses (anterior and posterior), located between the diaphragma sellae and the hypophyseal fossa

Which four cranial nerves are found within the lateral wall of the cavernous sinus?

1. CN III (the oculomotor nerve)
2. CN IV (the trochlear nerve)
3. CN V_1 (the ophthalmic division of the trigeminal nerve)
4. CN V_2 (the maxillary division of the trigeminal nerve)

Which nerve and which vessel run inside the cavernous sinus?	1. CN VI (the abducens nerve) 2. The ophthalmic artery, a branch of the internal carotid artery
What is an arteriovenous fistula?	A mixing of arterial and venous blood that occurs when trauma results in a tear of the internal carotid artery within the cavernous sinus
Which three veins drain into the cavernous sinus?	The superior and inferior ophthalmic veins, and the great cerebral vein (of Galen)
Which small sinus, lying in the margin of the tentorium cerebelli, runs from the cavernous sinus to the transverse sinus?	The superior petrosal sinus
Which small sinus drains the cavernous sinus into the internal jugular vein?	The inferior petrosal sinus
Describe the two courses venous blood may take between the cavernous sinus and the internal jugular vein.	1. Cavernous sinus, superior petrosal sinus, transverse sinus, sigmoid sinus, internal jugular vein 2. Cavernous sinus, inferior petrosal sinus, internal jugular vein
What is infectious inflammation of the cavernous sinus with secondary thrombus formation called?	Cavernous sinus thrombophlebitis
Which is the smallest of the venous sinuses?	The occipital sinus
Where is the occipital sinus located?	In the attached margin of the falx cerebelli
Where does the occipital sinus drain?	Into the "confluence of sinuses"

CEREBROSPINAL FLUID (CSF)

What is CSF?	CSF is the liquid that fills the subarachnoid space, surrounding and cushioning the brain and spinal cord

Where is CSF produced?

In the vascular choroid plexuses in the ventricles of the brain

Name the four ventricles.

1. Right lateral ventricle
2. Left lateral ventricle
3. Third ventricle
4. Fourth ventricle

What is the name of the thin wall that separates the right and left lateral ventricles?

The septum pellucidum

Which ventricle is contained within the diencephalon?

The third ventricle

Which ventricle lies between the pons and the cerebellum?

The fourth ventricle

Describe the course of CSF from the lateral ventricle to the fourth ventricle.

Lateral ventricle, interventricular foramen (of Monro), third ventricle, cerebral aqueduct (of Sylvius), fourth ventricle

Where is the cerebral aqueduct (of Sylvius) located?

In the midbrain

How does CSF travel from the fourth ventricle to the subarachnoid space?

From the fourth ventricle, CSF passes through the foramen of Magendie (medially) and the two foramina of Luschka (laterally) into the cerebello-medullary cistern (i.e., the cisterna magna), a large pool of CSF that widely separates the arachnoid mater and the pia mater.

Identify the subarachnoid cisterns on the following figure:

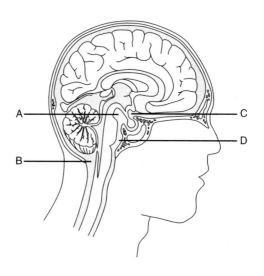

A = Interpeduncular cistern
B = Cerebellomedullary cistern (i.e., the cisterna magna)
C = Chiasmatic cistern
D = Pontine cistern

Which important blood vessel lies protected within the pontine cistern?

The basilar artery (remember, the pons and the basilar artery go together)

The posterior part of the circle of Willis is located within which cistern?

The interpeduncular cistern (between the cerebral peduncles)

How is CSF reabsorbed into the venous circulation?

Through the arachnoid villi, projections of the arachnoid mater that extend into the venous (dural) sinuses

What are aggregations of arachnoid villi called?

Arachnoid granulations

Where does most CSF pass into the venous blood?	At the superior sagittal sinus

THE SPINAL CORD

VERTEBRAL COLUMN

What structure supports and protects the spinal cord?	The vertebral column, comprised of vertebrae and intervertebral disks
What is a vertebral foramen?	The large central circular opening within each vertebra, bounded by lamina posteriorly, pedicles laterally, and the vertebral body anteriorly; in the articulated vertebral column, the vertebral foramina form the vertebral canal, which houses the spinal cord and its associated structures
How many vertebrae comprise the vertebral column?	33, although only 24 are movable in adults
What are the five regions of the vertebral column, and how many vertebrae comprise each?	Cervical: 7 Thoracic: 12 Lumbar: 5 Sacral: 5 (fused) Coccygeal: 4 (fused)
Where does the spinal cord end in:	
Adults?	Vertebral level L2
Newborns?	Vertebral level L3
What is the tapered (cone-shaped) lower end of the spinal cord called?	The conus medullaris

SPINAL NERVES

What is the distribution of the 31 pairs of spinal nerves that leave the spinal cord via the intervertebral foramina?	Cervical: 8 Thoracic: 12 Lumbar: 5 Sacral: 5 Coccygeal: 1
Where does spinal nerve C5 exit the vertebral column relative to vertebra C5?	Above it

Where does spinal nerve T5 exit the vertebral column relative to vertebra T5?	Below it
Explain why spinal nerve C5 exits above its associated vertebra, while spinal nerve T5 exits below.	This is a function of the fact that there are 8 cervical spinal nerves and only 7 cervical vertebra; spinal nerve C8 exits below vertebra C7 and above vertebra T1, forcing spinal nerve T1 to exit below vertebra T1, and so on.
What is the cauda equina ("horse's tail")?	The splayed bundle of elongated spinal nerve roots caudal to the termination of the spinal cord (i.e., below vertebra L2)
How are the spinal nerves formed?	Rootlets from the dorsal and ventral surfaces of the spinal cord unite to form the dorsal and ventral roots of the spinal nerve. Upon exiting the vertebral canal, the dorsal and ventral roots join together to form the spinal nerve.
What type of information does the dorsal root of a spinal nerve transmit, and in what direction?	Sensory ("afferent") information, toward the spinal cord
What type of information does the ventral root of a spinal nerve transmit, and in what direction?	Motor ("efferent") information, away from the spinal cord
Describe the location of the sensory dermatome associated with each of the following dorsal roots:	
C6	Radial forearm and thumb ("six-shooter")
T4	Nipple
T10	Umbilicus
S3	Genitoanal region
Which muscles are supplied by ventral roots from:	
C3–C5?	The diaphragm ("C3, C4, and C5 keep the diaphragm alive")

C8–T1?	Muscles of the hand
S1–S2?	Ankle plantar flexors
S3–S5?	Muscles of the bladder, anal sphincter, and genitals
What structures does the spinal nerve divide into, almost immediately upon exiting the vertebral canal?	The dorsal and ventral primary rami
Does the dorsal primary ramus transmit motor or sensory information? What about the ventral primary ramus?	Rami convey both sensory and motor information; roots convey one type or the other

SPINAL TRACTS

What is a spinal tract?	A bundle of axons (i.e., white matter) connecting parts of the CNS
Through which spinal tracts are tactile, vibratory, and proprioceptive (joint position) sense transmitted from the spinal cord to the brain?	The fasciculus gracilis and fasciculus cuneatus (ascending tracts of the dorsal column) and the dorsolateral tract of Lissauer
What type of sensory information does the spinothalamic tract transmit from the spinal cord to the thalamus?	Pain and temperature
Where is the spinothalamic tract located?	In the ventrolateral aspect of the spinal cord
Which spinal tract conveys motor information from the cerebral cortex to the spinal cord?	The corticospinal tract (also known as the "pyramidal tract")

VASCULATURE OF THE SPINAL CORD

What arteries constitute the arterial blood supply of the spinal cord?	1. The unpaired anterior spinal artery 2. Two posterior spinal arteries 3. The radicular ("root") branches of the

vertebral, cervical, posterior
intercostal, and lumbar arteries

**The anterior spinal artery
is a branch of which artery?**

The two vertebral arteries (they combine
short branches to become the anterior
spinal artery)

**Which radicular artery pro-
vides the main blood supply
to the inferior two thirds of
the spinal cord?**

The arteria radicularis magna, clinically
known as the "artery of Adamkiewicz"

MENINGES OF THE SPINAL CORD

**Name the meningeal layers
that surround the spinal
cord.**

The meninges that surround the spinal
cord are continuations of those that
surround the brain (i.e., the pia mater,
arachnoid mater, and dura mater).

**Which bony landmarks can
be used to determine the
location of vertebrae
L3–L4?**

The iliac crests are on a horizontal line
with vertebrae L3–L4.

What is the filum terminale?

The filum terminale, an extension of the
pia mater that attaches to the coccyx,
represents the most caudal extension of
the spinal cord tissue.

POWER REVIEW

**What is the name of a
collection of nerve cell
bodies located:**

    Within the CNS

Neuron

    Outside of the CNS?

Ganglion

BRAIN

**Which structure joins the
cerebral hemispheres?**

The corpus callosum

**What structures are separ-
ated from one another by
the lateral sulcus (i.e., the
sylvian fissure)?**

The lateral sulcus separates the temporal
lobe from the frontal and parietal lobes.

Which portion of the brain is concerned with:

Voluntary motor activity?	The precentral gyrus
Sensory data?	The postcentral gyrus

What are the three meninges, from superficial to deep?

Dura mater, arachnoid mater, pia mater

Name the four dural folds that subdivide the brain.

Falx cerebri, tentorium cerebelli, falx cerebelli, diaphragma sellae

What are the two midline dural folds?

The falx cerebri and the falx cerebelli

Describe the falx cerebri.

The falx cerebri is the dural fold located in the longitudinal fissure between the two cerebral hemispheres.

Describe the tentorium cerebelli.

The tentorium cerebelli is the horizontal dural fold that supports the occipital lobes and covers the cerebellum.

What is the term "supratentorial" used to refer to?

Structures located above the tentorium cerebelli

Each internal carotid artery ends by dividing into which two arteries?

The anterior cerebral artery and the middle cerebral artery

Through which intracranial artery does the majority of blood from the internal carotid artery flow?

The middle cerebral artery

The vertebral arteries join to form which structure?

The basilar artery

How does the basilar artery end?

By dividing into the left and right posterior cerebral arteries

How does blood travel from the great cerebral vein (of Galen) back to the heart?

After entering the straight sinus, the blood passes to the transverse sinus and then to the sigmoid sinus, which drains into the internal jugular vein. The internal jugular vein drains into the

brachiocephalic vein, which in turn
empties into the superior vena cava.

**Which spaces communicate
through the:**

    **Interventricular foramen
    (of Monroe)?**

The two lateral ventricles and the third
ventricle

    **Cerebral aqueduct (of
    Sylvius)?**

The third ventricle and the fourth
ventricle

    **Foramen of Magendie
    (median aperture)?**

The fourth ventricle and the cerebro-
medullary cistern (i.e., the cisterna
magna)

    **Foramina of Luschka
    (lateral apertures)?**

The fourth ventricle and the cerebro-
medullary cistern (i.e., the cisterna magna)

**What structure separates
the lateral ventricles?**

The septum pellucidum

SPINAL CORD

**What structure is housed
in the:**

    Vertebral canal?

The spinal cord

    Intervertebral foramina?

Individual spinal nerves

**State where each of the
following spinal nerves exits
the vertebral canal:**

    Spinal nerve C1

Above vertebra C1 (i.e., the atlas)

    Spinal nerve C8

Below vertebra C7

**Which artery that supplies
the upper portion of the
spinal cord is paired?**

The posterior artery (a branch of the
vertebral artery); the anterior artery is
unpaired

**Which artery constitutes
the principal arterial supply
of the inferior two thirds of
the spinal cord?**

The artery of Adamkiewicz

4   The Cranial Nerves

What quality makes a nerve a cranial nerve?	To be a cranial nerve, a nerve must pass through a foramen in the skull.
How many pairs of cranial nerves are there?	Twelve

Give the names of the twelve cranial nerves:

I	Olfactory nerve
II	Optic nerve
III	Oculomotor nerve
IV	Trochlear nerve
V	Trigeminal nerve
VI	Abducens nerve
VII	Facial nerve
VIII	Vestibulocochlear nerve
IX	Glossopharyngeal nerve
X	Vagus nerve
XI	Spinal accessory nerve
XII	Hypoglossal nerve

For each cranial nerve, state whether it carries sensory fibers, motor fibers, or both:

I (olfactory nerve)	Sensory
II (optic nerve)	Sensory

III (oculomotor nerve)	Motor
IV (trochlear nerve)	Motor
V (trigeminal nerve)	Both
VI (abducens nerve)	Motor
VII (facial nerve)	Both
VIII (vestibulocochlear nerve)	Sensory
IX (glossopharyngeal nerve)	Both
X (vagus nerve)	Both
XI (spinal accessory nerve)	Motor
XII (hypoglossal nerve)	Motor
	Some **S**ay **M**arry **M**oney, **B**ut **M**y **B**rother **S**ays **B**ig **B**rains **M**atter **M**ore

Which cranial nerves carry parasympathetic fibers?	1. CN III (oculomotor nerve) 2. CN VII (facial nerve) 3. CN IX (glossopharyngeal nerve) 4. CN X (vagus nerve)
Are these parasympathetic fibers pre- or post-ganglionic?	Preganglionic
Which cranial nerves carry preganglionic sympathetic fibers?	None
For each cranial nerve or group of cranial nerves, state where the nerve or group of nerves enters or exits the brain:	
I	Anterior forebrain
II	Diencephalon

III and IV	Midbrain
V	Pons
VI, VII, and VIII	Junction of the pons and the medulla
IX, X, XI (cranial root), and XII	Medulla

Where does the spinal root of CN XI (the spinal accessory nerve) originate?

On the superior aspect of the spinal cord

Cranial nerves exit which aspect of the brain stem?

The ventral (anterior) surface

What is the one exception?

CN IV (the trochlear nerve) exits on the dorsal surface.

Which two cranial nerves are not peripheral nervous tissue?

CN I (the olfactory nerve) and CN II (the optic nerve)

Which three cranial nerves do not originate on the brain stem?

1. CN I (the olfactory nerve)
2. CN II (the optic nerve)
3. CN XI (the spinal accessory nerve)

Identify each of the labeled structures on the following figure of the base of the brain:

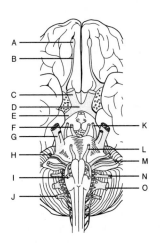

A = CN I (the olfactory nerve)
B = Olfactory tract
C = CN II (the optic nerve)
D = Optic chiasm
E = Optic tract
F = CN III (the oculomotor nerve)
G = CN V (the trigeminal nerve)
H = CN VII (the facial nerve)
I = CN XII (the hypoglossal nerve)
J = CN XI (the spinal accessory nerve)
K = CN IV (the trochlear nerve)
L = CN VI (the abducens nerve)
M = CN VIII (the vestibulocochlear nerve)
N = CN IX (the glossopharyngeal nerve)
O = CN X (the vagus nerve)

Identify each of the labeled cranial nerves on the following figure of the base of the skull:

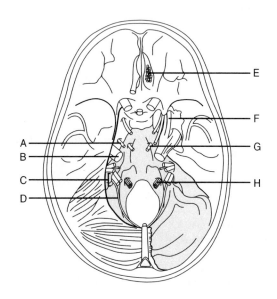

A = CN IV (the trochlear nerve)
B = CN V (the trigeminal nerve)
C = CN IX (the glossopharyngeal nerve), CN X (the vagus nerve), and CN XI (the spinal accessory nerve)
D = CN XI (the spinal accessory nerve)
E = CN I (the olfactory nerve)
F = CN V_1 (the opthalmic division of the trigeminal nerve)
G = CN VI (the abducens nerve)
H = CN XII (the hypoglossal nerve)

CN I (OLFACTORY NERVE)

What type of fibers are carried by CN I?

Special sensory fibers

What is the function of CN I?

Provides for the sense of smell

Describe the origin and course of the olfactory nerve.

Approximately 20 neurosensory cells unite in the superior nasal cavity to form the small nerve bundles that comprise the olfactory nerve. These bundles pass through foramina in the cribriform plate of the ethmoid bone to enter the olfactory bulbs in the anterior cranial fossa. From the olfactory bulbs, impulses are conveyed to cortical centers by the olfactory tracts.

What is the most common cause of anosmia (i.e., an inability to smell)?

Chronic rhinitis; other causes include fracture of the cribriform plate and tumors or abscesses of the frontal lobe that compress the olfactory bulb

CN II (OPTIC NERVE)

What type of fibers are carried by CN II?

Special sensory fibers

What is the function of CN II?

Vision

How does CN II enter the orbit?

Through the optic canal

Which structure accompanies the optic nerve in the optic canal?

The ophthalmic artery

Which terms are used to describe the parts of CN II in the visual pathway?

Optic nerve, optic chiasm, and optic tract

Where do the two optic nerves temporarily join together?

At the optic chiasm

The lateral retina becomes the optic tract on which side?

The ipsilateral side

Why?

These fibers do not cross at the optic chiasm.

The medial retina becomes the optic tract on which side?

The contralateral side

Why?

These fibers do cross at the optic chiasm.

Which three sheaths enclose the optic nerve?	The optic nerve is enclosed by the three layers of cerebral meninges.
The majority of the optic tract terminates in which structure?	The lateral geniculate body of the thalamus thalamus
The optic radiations connect which two structures?	The lateral geniculate body and the visual cortex

CN III (OCULOMOTOR NERVE)

Which types of fibers are carried by CN III?	Somatic and visceral motor fibers
What are the functions of CN III?	**Somatic motor fibers:** Innervate the medial, inferior, and superior recti muscles, the inferior oblique muscle, and the levator palpebrae superioris muscle **Visceral motor fibers:** Provide visceral motor (parasympathetic) innervation to the sphincter pupillae and ciliary muscles
CN III travels between which two arteries on the base of the brain?	The posterior cerebral and superior cerebellar arteries
CN III passes through the lateral wall of which sinus?	The cavernous sinus
CN III enters the orbit through which structure?	The superior orbital fissure
List three other structures that also pass through the superior orbital fissure.	1. CN IV (the trochlear nerve) 2. CN V₁ (the opthalmic division of the trigeminal nerve) 3. CN VI (the abducens nerve)
What are some common causes of CN III dysfunction?	1. Herniation of the brain through the foramen magnum (uncal herniation), as a result of increased intracranial pressure [caused by an expanding intracranial mass lesion or obstruction of cerebrospinal fluid (CSF) outflow] 2. Aneurysms 3. Fractures or inflammatory processes involving the cavernous sinus 4. Strokes 5. Multiple sclerosis

CN IV (TROCHLEAR NERVE)

Which type of fibers are carried by CN IV?

Somatic motor fibers

What is the function of CN IV?

Motor innervation of the superior oblique muscle, which moves the eyeball inferiorly and laterally (in isolation) and inferiorly (in conjunction with the inferior rectus muscle)

What is the trochlea?

A fibrocartilaginous structure attached to the frontal bone that serves as a pulley for the tendon of the superior oblique muscle

CN IV exits the skull through what opening?

The superior orbital fissure—remember, CN III (the oculomotor nerve), CN V_1 (the opthalmic division of the trigeminal nerve), and CN VI (the abducens nerve) also pass through this fissure!

List three unique features of CN IV.

1. It is the only cranial nerve to exit the brain stem on the dorsal surface.
2. It is the smallest of the cranial nerves in diameter.
3. It is the only cranial nerve whose peripheral fibers decussate before leaving the brain stem.

CN V (TRIGEMINAL NERVE)

CN V has three main divisions. Identify the territory of each on the following figure:

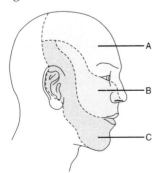

A = CN V_1 (the ophthalmic division)
B = CN V_2 (the maxillary division)
C = CN V_3 (the mandibular division)

Which part of the face is not innervated by CN V?	The angle of the mandible (innervated by spinal nerves C2 and C3)
Which type of fibers are carried by the trigeminal nerve?	Somatic sensory and branchial motor fibers
Which division of CN V carries all of the motor fibers?	CN V$_3$

CN V$_3$ provides motor innervation to which muscles?

1. The muscles of mastication (i.e., the medial and lateral pterygoids, the masseter muscle, and the temporalis muscle)
2. The tensor tympani muscle
3. The tensor veli palatini muscle
4. The mylohyoid muscle
5. The anterior belly of the digastric muscle

Where does each division of CN V exit the middle cranial fossa?

CN V$_1$: The superior orbital fissure
CN V$_2$: The foramen rotundum
CN V$_3$: The foramen ovale

CN VI (ABDUCENS NERVE)

CN VI carries which type of fibers?	Somatic motor
What is the function of CN VI?	The *abducens* nerve provides motor innervation to the lateral rectus muscle (which *abducts* the eye).
CN VI exits the skull through which opening?	The superior orbital fissure (remember, CN III, CN IV, and CN V$_1$ also pass through this fissure)

CN VII (FACIAL NERVE)

Which types of fibers are carried by the facial nerve?	Somatic sensory, visceral sensory, visceral motor (parasympathetic), and branchial motor fibers

Which muscles receive motor innervation from CN VII?

1. The muscles of facial expression
2. The stapedius muscle
3. The stylohyoid muscle

4. The posterior belly of the digastric
   muscle
   (Note the symmetry with CN V$_3$,
   which innervates the muscles of
   mastication, the tensor tympani
   muscle, the tensor veli palatini muscle,
   the mylohyoid muscle, and the
   anterior belly of the digastric muscle.)

**What are the sensory
functions of CN VII?**

Visceral (special) sensory functions:
   Taste (anterior two thirds of tongue)
   and sensory innervation of the soft and
   hard palates
Somatic sensory functions: Sensory
   innervation of the auricle and a small
   portion of skin behind the ear

**Which structures receive
parasympathetic innervation
from CN VII?**

1. The submandibular and sublingual
   salivary glands (via the chorda tympani
   branch)
2. The lacrimal glands (via the chorda
   tympani branch)
3. The secretory glands of the nasal and
   palatine mucosa (via the greater
   petrosal branch)

**What are the only two sets
of glands in the head that
CN VII does not innervate?**

1. The parotid gland (note that although
   CN VII passes through the parotid
   gland, it does not innervate it)
2. The integumentary glands (i.e., those
   of the scalp)

**Trace the path of CN VII
from the brain stem to its
exit from the skull.**

CN VII emerges from the caudal pons,
courses through the internal auditory
meatus with CN VIII (the
vestibulocochlear nerve), continues
through the facial canal in the petrous
portion of the temporal bone, and passes
through the stylomastoid foramen to exit
the cranial fossa.

**What is the greater petrosal
nerve?**

A branch of CN VII that carries taste
fibers to the tongue and parasympathetic fibers to the nasal and palatine
mucosa

**Describe the course of the
greater petrosal nerve.**

After originating from the geniculate
ganglion, the greater petrosal nerve

passes through the horizontal part of the facial canal and over the foramen lacerum. It then joins the deep petrosal nerve to form the nerve of the pterygoid canal, which passes through the pterygoid canal and then contributes to the pterygopalatine ganglion.

Where does the chorda tympani branch away from CN VII?

In the descending part of the facial canal

What is the function of this branch?

1. Provides taste to the anterior two thirds of the tongue
2. Provides parasympathetic innervation to the submandibular and sublingual salivary glands and the lacrimal glands

The chorda tympani exits the skull via which foramen?

The petrotympanic fissure

Which nerve does the chorda tympani join after exiting the petrotympanic fissure?

The lingual nerve, a branch of CN V_3

Lesions of the facial nerve proximal to the stylomastoid foramen lead to which deficits? Why?

1. Unilateral facial paralysis (loss of facial muscle innervation, also known as Bell's palsy)
2. Unilateral loss of taste in the anterior two thirds of the tongue (loss of chorda tympani taste fibers)
3. Decreased salivation (loss of innervation to the submandibular and sublingual glands)
4. Unilateral hyperacusis (loss of innervation to the stapedius muscle of the inner ear, which normally dampens sounds)

Lesions of the facial nerve distal to the stylomastoid foramen lead to what deficits?

Unilateral facial paralysis only. Taste, salivation, and hearing are spared because the branches that innervate the tongue, salivary glands, and stapedius muscle arise proximal to this point.

CN VIII (VESTIBULOCOCHLEAR NERVE)

CN VIII carries which type of fibers?

Special sensory fibers (like CN I and CN II)

What is the function of the vestibulocochlear nerve?

Hearing (cochlear portion) and balance (vestibular portion)

Where does CN VIII exit the brain stem?

Branches representing the vestibular and cochlear portions emerge separately at a groove between the pons and medulla.

Which organs are innervated by the vestibular branch?

The utricle, saccule, and semicircular canals (i.e., the three organs responsible for maintaining equilibrium)

In addition to balance, what else does the vestibular branch of CN VIII mediate?

Coordination of head and eye movements

Which organ is innervated by the cochlear branch?

The cochlea

Where does CN VIII exit the posterior cranial fossa?

Through the internal auditory meatus (in the petrous portion of the temporal bone)

Which other structures travel with CN VIII through the internal auditory meatus?

CN VII (i.e., the facial nerve) and the labyrinthine (internal acoustic) artery, a branch from the basilar artery

CN IX (GLOSSOPHARYNGEAL NERVE)

Which types of fibers are carried by CN IX?

CN IX carries all types of cranial nerve fibers: visceral and branchial motor fibers, somatic fibers, visceral and special sensory fibers, and parasympathetic fibers.

What are the motor functions of CN IX?

Branchial motor fibers: Innervate the stylopharyngeus muscle, which elevates the pharynx during swallowing and speech
Visceral motor fibers: Supply the otic ganglion, which provides secretomotor fibers to the parotid gland

What are the sensory functions of CN IX?

Somatic sensory fibers: Provide sensation to the upper pharynx, tonsils, posterior third of the tongue, skin of the external ear, and the internal portion of the tympanic membrane

Special sensory fibers: Provide the posterior third of the tongue with taste sensation

Visceral sensory fibers: Carry afferent input from the carotid body and sinus

Which structure receives parasympathetic innervation from CN IX?

The parotid gland (remember—the parotid gland is traversed by CN VII but innervated by CN IX)

Where does CN IX exit the cranium?

At the jugular foramen (located at the suture line between the inferior edges of the temporal and occipital bones)

Which other cranial nerves exit the skull at the jugular foramen?

CN X (the vagus nerve) and CN XI (the spinal accessory nerve)

What are the six major branches of CN IX?

1. Tympanic branch
2. Carotid branch
3. Pharyngeal branch
4. Muscular branch
5. Tonsillar branch
6. Lingual branch

What is the key to finding CN IX on a dissection?

Locate the stylopharyngeus muscle; CN IX emerges posteriorly and runs around the lateral border of the muscle.

CN X (VAGUS NERVE)

Which types of fibers does CN X carry?

Branchial and visceral motor fibers and somatic and visceral sensory fibers

What are the motor functions of CN X?

Branchial motor fibers: Innervate the striated muscles of the pharynx and larynx (except for the stylopharyngeus and tensor veli palatini muscles), and the palatoglossus muscle of the tongue

Visceral motor fibers: Innervate the smooth muscle of the abdominal and thoracic viscera

What are the sensory functions of CN X?

Somatic sensory fibers: Innervate the skin on the back of the ear, the external auditory meatus, the external tympanic membrane, and the pharynx

Visceral sensory fibers: Innervate the larynx, trachea, and esophagus; the thoracic and abdominal viscera; the stretch receptors in the aortic arch, and the chemoreceptors in the aortic bodies

Which structures receive parasympathetic innervation via CN X?

1. The cardiac plexus (parasympathetic innervation slows the heart and constricts coronary arteries)
2. The pulmonary plexus (parasympathetic innervation constricts the bronchial tree)
3. The abdominal branches (parasympathetic innervation provides for gastrointestinal motility as far as the left colic flexure and stimulates some gastrointestinal secretions)

Describe the origin of CN X.

CN X arises as 8–10 rootlets from the medulla.

List the seven major branches of CN X.

1. Meningeal branch (provides sensation to the dura mater in the posterior cranial fossa)
2. Auricular branch (provides sensation to the back of the ear and communicates with the auricular branch of CN VII)
3. Superior laryngeal branch (the internal laryngeal branch provides sensation in the larynx above the vocal cords; the external laryngeal branch provides motor innervation to the inferior constrictor and cricothyroid muscles)
4. Recurrent laryngeal branch (innervates all of the intrinsic muscles of the larynx except the cricothyroid muscle)
5. Nerve to the carotid body and sinus
6. Motor branch to the pharyngeal plexus
7. Parasympathetic branches to the thoracic and abdominal viscera

Where does CN X travel in the neck?

Within the carotid sheath, posterior to the carotid artery and medial to the internal jugular vein

What structure does the recurrent laryngeal nerve loop around on the:

Right side?

The subclavian artery

Left side?

The ligamentum arteriosum

The vagus joins with which structure before leaving the thorax?

The contralateral vagus (forming the esophageal plexus and vagal trunks)

Which vagus nerve (right or left) forms the anterior vagal trunk?

The left
(Remember **LARP**: **L**eft **A**nterior,
    Right **P**osterior)

How can the variation in the paths of the left and right vagus nerves be explained?

By recalling the embryologic development of the region; as the foregut rotates (clockwise from above), the vagus nerve (which is adherent to the esophagus) rotates with it.

List the six structures supplied by the anterior vagal trunk.

1. The anterior aspect of the stomach
2. Lesser omentum
3. Liver
4. Pylorus
5. Head of the pancreas
6. The first two parts of the duodenum

Which structure is supplied by the posterior vagal trunk?

The posterior aspect of the stomach

CN XI (SPINAL ACCESSORY NERVE)

CN XI carries which type of fibers?

Somatic motor fibers

What is the function of CN XI?

Spinal root: Provides motor innervation to the sternocleidomastoid and trapezius muscles

Cranial root: Provides somatic motor innervation to the larynx and pharynx via the pharyngeal and recurrent laryngeal branches of CN X

Describe the course of the spinal root of CN XI.

After originating from cervical segments C1–C6, the spinal root of CN XI travels superiorly into the cranium through the foramen magnum, joins the cranial root, and then exits the skull through the jugular foramen.

CN XII (HYPOGLOSSAL NERVE)

CN XII carries which type of fibers?	Somatic motor fibers
What is the function of CN XII?	CN XII provides motor innervation to all of the intrinsic and extrinsic muscles of the tongue, except for the palatoglossus muscle (which is supplied by CN X).
Where does CN XII originate?	CN XII emerges from the brain stem as 8–10 rootlets between the "olive" and the pyramid of the ventral medulla.
Where does CN XII exit the skull?	Through the hypoglossal canal (in the occipital bone)
Does CN XII travel medial or lateral to CNs IX, X, and XI upon exiting the cranium?	Medial (think of the tongue being in the middle)
Which spinal nerve fibers travel with CN XII?	The descending branches of spinal nerve C1, which join with branches from spinal nerves C2 and C3 to form the ansa cervicalis

POWER REVIEW

For each cranial nerve, state the cranial foramen through which it passes:	
I (the olfactory nerve)	The cribriform plate
II (the optic nerve)	The optic canal
III (the oculomotor nerve)	The superior orbital fissure
IV (the trochlear nerve)	The superior orbital fissure
V₁ (the opthalmic division of the trigeminal nerve)	The superior orbital fissure

V_2 (**the maxillary division of the trigeminal nerve**)	The foramen rotundum
V_3 (**the mandibular division of the trigeminal nerve**)	The foramen ovale
VI (the abducens nerve)	The superior orbital fissure
VII (the facial nerve)	The stylomastoid foramen
VIII (the vestibulo-cochlear nerve)	The internal auditory meatus
IX (the glossopharyngeal nerve)	The jugular foramen
X (the vagus nerve)	The jugular foramen
XI (the spinal accessory nerve)	The jugular foramen
XII (the hypoglossal nerve)	The hypoglossal canal
Which four cranial nerves carry parasympathetic fibers?	1. CN III (the oculomotor nerve) 2. CN VII (the facial nerve) 3. CN IX (the glossopharyngeal nerve) 4. CN X (the vagus nerve)
Where do the two optic nerves temporarily join together?	At the optic chiasm
Between which two arteries does CN III (the oculomotor nerve) emerge from the brain?	CN III emerges from the brain between the superior cerebellar and posterior cerebral arteries.
What is the function of CN IV (the trochlear nerve)?	Motor innervation to the superior oblique muscle
What are the three main sensory divisions of CN V (the trigeminal nerve)?	1. CN V_1: The ophthalmic division 2. CN V_2: The maxillary division 3. CN V_3: The mandibular division
CN V provides motor innervation to which muscles?	1. The muscles of mastication 2. The tensor tympani muscle

3. The tensor veli palatini muscle
4. The mylohyoid muscle
5. The anterior belly of the digastric muscle

What is the function of CN VI (the abducens nerve)?

Motor innervation to the lateral rectus muscle, which abducts the eye

CN VII (the facial nerve) provides motor innervation for which structures?

1. The muscles of facial expression
2. The stapedius muscle
3. The stylohyoid muscle
4. The posterior belly of the digastric muscle

Which nerve provides for taste on the:

Anterior two thirds of the tongue?

The chorda tympani, a branch of CN VII

Posterior third of the tongue?

CN IX (the glossopharyngeal nerve)

What is the function of CN VIII (the vestibulo-cochlear nerve)?

Provides for hearing and equilibrium

What are the symptoms and signs of vestibular nerve dysfunction?

Dizziness, impaired balance, nystagmus, and nausea or vomiting

Where does CN X (the vagus nerve) travel in the neck?

Within the carotid sheath, posterior to the carotid artery and medial to the internal jugular vein

The recurrent laryngeal nerve loops around which structure on the right? On the left?

The subclavian artery and the ligamentum arteriosum, respectively (remember, unilateral damage causes hoarseness; bilateral damage causes airway obstruction)

Which vagus nerve (i.e., the right or the left) forms the anterior vagal trunk?

The left
(Remember **LARP: L**eft **A**nterior, **R**ight **P**osterior)

Which intrinsic muscle of the larynx is not supplied by the recurrent laryngeal nerve?

The cricothyroid muscle

What is the function of CN XI (the spinal accessory nerve)?

The spinal root supplies motor innervation to the sternocleidomastoid and trapezius muscles, and the cranial root supplies motor innervation to the larynx and pharynx.

What is the function of CN XII (the hypoglossal nerve)?

Moves the tongue (if CN XII is damaged, the tongue will deviate toward the side of the lesion)

5   The Neck

BONES OF THE NECK

Which bones form the skeleton of the neck?

The seven cervical vertebrae (see Chapter 6, "The Back") and the clavicles (see Chapter 7, "The Upper Limb").

What are the superior bony landmarks of the neck?

The inferior margin of the mandible, the mastoid process of the temporal bone, and the external occipital protuberance of the occipital bone

What are the inferior bony landmarks of the neck?

The superior borders of the clavicles and the manubrium (i.e., the first bone of the sternum)

Which small, "U"-shaped bone, located at the level of the body of vertebra C3 and just below the mandible, serves as an attachment site for many of the muscles of the neck?

The hyoid bone

Which two muscles originate on the hyoid bone?

The hyoglossus (one of the extrinsic tongue muscles) and the middle constrictor muscle of the pharynx

List the seven muscles of the neck that insert on the hyoid bone.

1. The mylohyoid muscle
2. The geniohyoid muscle
3. The stylohyoid muscle
4. The digastric muscle
5. The omohyoid muscle
6. The sternohyoid muscle
7. The thyrohyoid muscle

MUSCLES OF THE NECK

SUPERFICIAL MUSCLES

What are the two superficial muscles of the neck?

1. The sternocleidomastoid muscle
2. The platysma

As you make a transverse cut across the neck to do a thyroid resection, what is the most superficial muscle you encounter?

The platysma

Platysma

What are the attachments of the platysma muscle?

This thin, flat muscle extends from the mandible to the fascia of the pectoralis and deltoid muscles of the upper extremity. (*Plat-* means "flat.")

What is the innervation of the platysma?

The platysma is innervated by the cervical branch of CN VII (the facial nerve).

What does the platysma do?

Assists in facial expressions, especially frowning

Sternocleidomastoid muscle

Origin?

The mastoid process of the skull

Insertions?

As the sternocleidomastoid muscle runs anteromedially, it splits, forming a sternal head (which inserts on the sternum) and a clavicular (cleido-) head, which inserts on the clavicle.

Innervation?

The spinal branches of CN XI (the accessory spinal nerve) and spinal nerves C2 and C3

Actions?

Acting singly, the sternocleidomastoid muscle pulls on the mastoid process, tilting the head to the ipsilateral side, laterally bending the neck, and rotating the face so that it looks superiorly to the other side. Acting together, the sternocleidomastoid muscles flex the neck.

DEEP MUSCLES

Which three major categories are used to classify the deep muscles of the neck?

1. Suprahyoid group
2. Infrahyoid group (strap muscles)
3. Muscles of the posterior triangle floor

Suprahyoid muscles

Which four muscles comprise the suprahyoid group?

1. The mylohyoid
2. The geniohyoid
3. The stylohyoid
4. The digastric

Mylohyoid muscle

Origin?   The mylohyoid line of the mandible

Insertion?   The median raphe of the hyoid bone

Innervation?   The mylohyoid nerve, a branch of CN V_3 (the mandibular branch of the trigeminal nerve)

Action?   Elevates the hyoid bone and tongue and depresses the mandible

Geniohyoid muscle

Origin?   The genial tubercle of the mandible

Insertion?   The body of the hyoid bone

Innervation?   Spinal nerve C1, via CN VII (the hypoglossal nerve)

Action?   Elevates the hyoid bone and tongue

Stylohyoid muscle

Origin?   The styloid process of the temporal bone

Insertion?   The body of the hyoid bone

Innervation?   CN VII (the facial nerve)

Action?   Elevates the hyoid bone

Digastric muscle

What is the origin of the digastric muscle's name?   The digastric muscle has an anterior belly and a posterior belly. "Digastric" means "two bellies."

Describe the course of the digastric muscle.	The anterior belly arises from the mandible and the posterior belly arises from the temporal bone, just deep to the mastoid process. The two bellies are connected by a tendon held in a fascial sling attached to the hyoid bone.
Which two nerves innervate the digastric muscle?	**Anterior belly:** The mylohyoid nerve, a branch of the inferior alveolar nerve (from CN V_3) **Posterior belly:** CN VII (the facial nerve)

Infrahyoid (strap) muscles

Which four muscles comprise the infrahyoid group?	1. The omohyoid 2. The sternohyoid 3. The sternothyroid 4. The thyrohyoid
What is the action of the strap muscles?	These muscles lower the hyoid and larynx during phonation.
What nerve innervates all of the infrahyoid muscles but one?	The ansa cervicalis
Which infrahyoid muscle is not innervated by the ansa cervicalis?	The thyrohyoid muscle—this muscle is innervated by spinal nerve C1 via CN XIII (the hypoglossal nerve)
Which strap muscle crosses perpendicular and just deep to the sternocleidomastoid muscle?	The omohyoid muscle
Which muscle has both a superior and an inferior belly?	The omohyoid muscle. After originating at the hyoid bone, the omohyoid muscle passes through a fascial sling (connected to the clavicle) and under the sternocleidomastoid muscle to insert on the superior border of the scapula. Therefore, it has both a superior and an inferior belly.

Omohyoid muscle

Origin?	Hyoid bone
Insertion?	Scapula (*omo*-means "shoulder" in Greek)

Sternohyoid muscle

Origin?   Manubrium of the sternum

Insertion?   Hyoid bone

Sternothyroid muscle

Origin?   Manubrium of the sternum

Insertion?   The hyoid bone

Thyrohyoid muscle

Origin?   The thyroid cartilage

Insertion?   The hyoid bone

Muscles of the posterior triangle floor

Which four deep muscles   1. Splenius capitis muscle
form the floor of the   2. Levator scapulae muscle
posterior triangle?   3. Posterior scalene muscle
                                     4. Middle scalene muscle

Which other muscle may   The anterior scalene muscle
contribute to the infero-
medial part of the posterior
triangle?

Where does the anterior   The first rib
scalene muscle insert?

Identify the muscles of the
neck on the following figure:

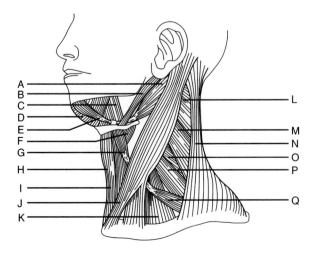

A = Digastric muscle, posterior belly
B = Stylohyoid muscle
C = Hyoglossus muscle
D = Mylohyoid muscle
E = Digastric muscle, anterior belly
F = Thyrohyoid muscle
G = Omohyoid muscle, superior belly
H = Sternothyroid muscle
I = Sternohyoid muscle
J = Sternocleidomastoid muscle
K = Anterior scalene muscle
L = Splenius capitis muscle
M = Levator scapulae muscle
N = Trapezius muscle
O = Posterior scalene muscle
P = Middle scalene muscle
Q = Omohyoid muscle, inferior belly

Splenius capitis muscle

Origin?

The inferior half of the nuchal ligament and the spinous processes of vertebrae C1–C6

Insertion?

The mastoid process and the lateral superior nuchal line

Action?

Unilateral contraction: Flexes and rotates the head and neck ipsilaterally
Bilateral contraction: Extends the head and neck

Levator scapulae muscle

Origin?

The transverse processes of vertebrae C1–C4

Insertion?

The superomedial border of the scapula

Innervation?

The dorsal scapular nerves and spinal nerves C3 and C4

Action?

Elevates and rotates the scapula (counterclockwise from the back)

Posterior scalene muscle

Origin?

The transverse processes of vertebrae C4–C6

Insertion?	The second rib
Innervation?	Spinal nerves C7 and C8
Action?	Flexes the neck laterally and elevates the second rib during inspiration

Middle scalene muscle

Origin?	The transverse processes of vertebrae C2–C7
Insertion?	The first rib
Innervation?	Spinal nerves C3–C8
Action?	Flexes the neck laterally and elevates the first rib during inspiration

FASCIAE OF THE NECK

SUPERFICIAL FASCIA

What is the superficial fascia of the neck called?	The investing fascia
What does the investing fascia invest?	The investing fascia is found on both sides of (i.e., it "invests") the sternocleidomastoid muscle. It runs inferior to the platysma and surrounds all of the deeper structures of the neck.

DEEP FASCIAE

What are the three deep fasciae of the neck?	1. Pretracheal fascia 2. Prevertebral fascia 3. Fascia of the carotid sheath
What does the pretracheal fascia enclose?	Located deep to the infrahyoid muscles, this fascia surrounds the trachea, thyroid glands, and esophagus.
What does the inferior border of the pretracheal fascia merge with?	The fibrous pericardium of the mediastinum
What does the prevertebral fascia cover?	Located posterior to the esophagus and pretracheal fascia, this fascia lies against

the cervical vertebral bodies and surrounds the prevertebral muscles.

What is the retropharyngeal space?

The retropharyngeal space is a potential space between the prevertebral fascia and the posterior portion of the pretracheal fascia.

TRIANGLES OF THE NECK

Which structures delineate the anterior triangle of the neck?

Anterior boundary: Median line of the neck
Posterior boundary: The anterior border of the sternocleidomastoid muscle
Base: Inferior mandible
Apex: Jugular notch (i.e., the space between the clavicles and above the manubrium)

Which two muscles divide the anterior triangle into four smaller triangles?

The digastric and omohyoid muscles

What are the four sub-divisions of the anterior triangle?

1. The submental triangle
2. The digastric (submandibular) triangle
3. The carotid triangle
4. The muscular triangle

Which structures delineate the posterior triangle of the neck?

Anterior boundary: The posterior border of the sternocleidomastoid muscle
Posterior boundary: The anterior border of the trapezius muscle
Base: The middle third of the clavicle
Apex: Point where the sternocleidomastoid and trapezius muscles meet, on the occipital bone

Which structure divides the posterior triangle into two smaller triangles, and what are the resultant spaces called?

The inferior belly of the omohyoid muscle divides the posterior triangle into the occipital and supraclavicular triangles.

Identify the labeled structures and triangles on the following illustration:

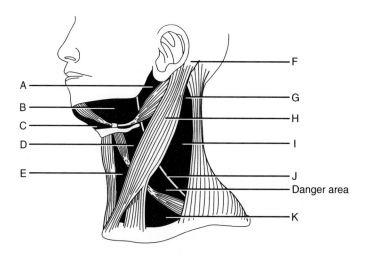

A = Parotid region
B = Digastric (submandibular) triangle
C = Submental triangle
D = Carotid triangle
E = Muscular triangle
F = Mastoid process
G = Apex of the posterior triangle
H = Sternocleidomastoid muscle
I = Occipital triangle
J = CN XI (the spinal accessory nerve)
K = Supraclavicular triangle

ANTERIOR TRIANGLE

Submental triangle

What is unique about the submental triangle?

It is the only unpaired triangle within the anterior triangle.

What are its boundaries?

The small submental triangle (*mental*, "chin") is bounded by the hyoid bone and the anterior bellies of the left and right digastric muscles, which form the apex of the triangle at the mandible.

What forms the floor of the submental triangle?

The mylohyoid muscle

| **Which structures of interest are found in the submental triangle?** | The submental lymph nodes, which drain the tip of the tongue, the lower incisors, and the lower lip and chin |

Digastric (submandibular) triangle

What bounds the digastric triangle?	The inferior margin of the mandible and the superior margins of both the anterior and posterior bellies of the digastric muscle
What structure nearly fills the digastric triangle?	The submandibular gland
Which major nerve passes through the digastric triangle?	CN XII (the hypoglossal nerve)

Carotid triangle

| **What are the boundaries of the carotid triangle?** | The anterior border of the sternocleidomastoid muscle, the superior belly of the omohyoid muscle, and the posterior belly of the digastric muscle |
| **Which important structures pass through the carotid triangle?** | The common internal artery, the internal jugular vein, and the vagus nerve (CN X), housed in the carotid sheath (occasionally, the superior part of the ansa cervicalis descends within the sheath as well) |

Identify the following structures in the carotid triangle:

A = Digastric muscle, posterior belly
B = CN XI (spinal accessory nerve)
C = Internal jugular vein
D = Carotid sheath
E = Internal carotid artery
F = External carotid artery
G = Facial artery
H = Hyoglossus muscle
I = Mylohyoid muscle
J = Lingual artery
K = CN XII (hypoglossal nerve)
L = Common carotid artery
M = CN X (vagus nerve)

Muscular triangle

What are the boundaries of the muscular triangle?	The anterior border of the sternocleidomastoid muscle and the inferior border of the superior belly of omohyoid
What divides the muscular triangle into right and left halves?	The midline of the neck
What are the principle contents of the muscular triangle?	The infrahyoid muscles, the thyroid gland, and the parathyroid glands

VASCULATURE OF THE NECK

ARTERIES

Where do the arteries of the neck arise from?	The arch of the aorta
What are the arteries of the neck:	
On the left side?	1. The left common carotid artery 2. The left subclavian artery
On the right side?	The brachiocephalic trunk (innominate artery)
Which two arteries does the brachiocephalic trunk divide into?	1. The right common carotid artery 2. The right subclavian artery

Subclavian artery

How does the subclavian artery pass out of the thorax and into the neck?

Over the first rib, between the anterior and middle scalenus muscles, and under the clavicle

Which structure divides the subclavian artery, and into what parts?

The subclavian artery is divided into three parts as it passes behind the anterior scalenus muscle. The first part is medial to the muscle, the second part is posterior to it, and the third part is lateral to it.

Which nerve loops under the right subclavian artery?

The right recurrent laryngeal nerve, a branch of CN X (the vagus nerve)

What are the five major branches off the subclavian artery?

1. The internal thoracic (internal mammary) artery
2. The vertebral artery
3. The thyrocervical trunk
4. The costocervical trunk
5. The dorsal scapular artery

With one exception, which part of the subclavian artery sends off all of the branches? What is the exception?

All branches arise from the first part of the subclavian artery, except for the costocervical trunk on the right, which usually arises from the second part.

What do the vertebral arteries become?

The vertebral arteries travel in the vertebral foramina of the cervical vertebrae and join posterior to the pons to form the basilar artery. (Recall that the vertebrobasilar system, along with the internal carotid system, provides blood to the brain.)

What are the three branches of the thyrocervical trunk?

1. The inferior thyroid artery
2. The transverse cervical artery
3. The suprascapular artery

What does the costocervical trunk divide into?

The superior intercostal and deep cervical arteries

What does the subclavian artery become as it leaves the neck and travels distally? When does this occur?

The subclavian artery becomes the axillary artery as it passes the lateral border of the first rib.

Common carotid artery

How does the common carotid artery terminate?

By dividing into the internal and external carotid arteries

Where does the common carotid bifurcate?

Within the carotid triangle, at the superior border of the thyroid cartilage

What is the difference between the carotid body and the carotid sinus, in terms of structure, function, and location?

The carotid body is a small, ovoid mass of tissue located at the bifurcation of the common carotid artery into the internal and external carotid arteries. The carotid body contains chemoreceptors that detect changes in oxygen and carbon dioxide tensions. The carotid sinus is a dilatation of the internal carotid artery just distal to the bifurcation that contains pressure receptors.

Which nerve innervates the carotid sinus?

CN IX (the glossopharyngeal nerve)

What are the branches of the internal carotid artery within the neck?

There are none! (The first branch of the internal carotid artery is the ophthalmic artery, which arises intracranially.

What are the two terminal branches of the external carotid artery?

The maxillary artery and the superficial temporal artery

What are the six branches of the external carotid artery before this bifurcation occurs, in ascending order?

1. Superior thyroid artery
2. Ascending pharyngeal artery
3. Lingual artery
4. Facial artery
5. Occipital artery
6. Posterior auricular artery

Of these six branches, which are:

Anterior?

1. The superior thyroid artery
2. The lingual artery
3. The facial artery
(Arguably the most important three)

Posterior?

1. The occipital artery
2. The posterior auricular artery

Medial?

The ascending pharyngeal artery

Identify the arteries of the
neck on the following figure:

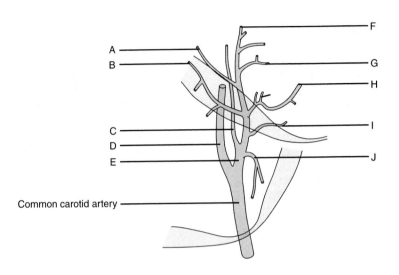

A = Posterior auricular artery
B = Occipital artery
C = Ascending pharyngeal artery
D = Internal carotid artery
E = External carotid artery
F = Superficial temporal artery
G = Maxillary artery
H = Facial artery
I = Lingual artery
J = Superior thyroid artery

VEINS

Internal jugular vein

What areas are drained by
the internal jugular vein?

Structures in the cranium and the face

Where does the internal
jugular vein originate?

A direct continuation of the sigmoid
sinus, the internal jugular vein exits the
skull through the jugular foramen.

Which fascia covers the
internal jugular vein?

The carotid sheath (which also invests the
carotid artery and CN X)

What is the relation between the three structures within the carotid sheath?	The vein is lateral, the artery is medial, and the nerve is posterior.

External jugular vein

Which veins join to form the external jugular vein?	The retromandibular and posterior auricular veins
Into which structure does the external jugular vein drain?	The external jugular vein travels behind the sternocleidomastoid muscle to join the subclavian vein.

INNERVATION OF THE NECK

Which sympathetic nerves travel in the neck?	The sympathetic trunk contains postganglionic fibers which originate in the three cervical ganglia (i.e., the inferior, middle, and superior ganglia). These postganglionic sympathetic fibers travel up the neck along the carotid sheath.
Is the sympathetic trunk in the carotid sheath?	No, the sympathetic trunk travels alongside (outside and posterior to) the carotid sheath.
Interruption of the sympathetic trunk in the neck causes which syndrome?	Horner's syndrome
What is Horner's syndrome?	**Ptosis:** Drooping of the eyelid occurs when innervation of the levator palpebrae is disrupted. **Miosis:** The pupil is constricted owing to unopposed parasympathetic action. **Anhydrosis:** Perspiring is a sympathetic function.
Which lung tumor can cause a Horner's syndrome?	A superior sulcus tumor (Pancoast tumor)
Where do the parasympathetic nerves in the neck travel?	The carotid sheath—CN X (the vagus nerve) represents the parasympathetic fibers in the neck
Which important nerve travels with (and is part of) CN X?	The recurrent laryngeal nerve

Describe the path of the right recurrent laryngeal nerve.

After traveling down the neck as part of CN X, the right recurrent laryngeal nerve hoods around the subclavian artery to travel back up toward the larynx in the tracheoesophageal groove, while CN X passes anterior to the subclavian artery.

Describe the path of the left recurrent laryngeal nerve.

The left recurrent laryngeal nerve takes a similar path, except it hooks around the ligamentum arteriosum while CN X passes anterior to the arch of the aorta.

The cervical plexus originates from which nerves?

The ventral primary rami of spinal nerves C1–C4

Which motor branches originate from the cervical plexus?

1. The ansa cervicalis
2. The phrenic nerve
3. Twigs to the longus capitis and cervicis muscles, the sternocleidomastoid muscle, the trapezius muscle, the levator scapulae muscle, and the scalene muscles

Which nerve innervates the strap muscles?

The ansa cervicalis

What is the origin and course of this nerve?

This "U"-shaped nerve (*ansa* means "loop") originates from spinal nerves C1–C3 and runs just anterior to the carotid sheath. This position makes the ansa cervicalis vulnerable to injury during surgery on the carotid artery.

Which nerve innervates the diaphragm? What is its origin?

The phrenic nerve, which arises from spinal nerves C3–C5 ("C3, C4, and C5 keep the diaphragm alive")

What is the path of the phrenic nerve in the neck?

The phrenic nerve crosses the anterior of the anterior scalenus muscle to enter the thoracic inlet

What are the four major sensory branches of the cervical plexus?

1. The lesser occipital nerve
2. The great auricular nerve
3. The transverse cervical nerve
4. The supraclavicular nerve

Identify the labeled structures on the following figure:

A = Great auricular nerve
B = Transverse cervical nerve
C = Sternocleidomastoid muscle
D = Lesser occipital nerve
E = CN XI (spinal accessory nerve)
F = Trapezius muscle
G = Supraclavicular nerve

Between which muscles does the brachial plexus emerge from the deep part of the neck?

The anterior and medial scalenus muscles

VISCERA OF THE NECK

THYROID GLAND

What connects the left and right lobes of the thyroid gland?

The isthmus

What is the isthmus anterior to?

The second and third tracheal rings

Describe the origin of the dual blood supply of the thyroid.

The superior thyroid artery originates from the external carotid artery. The inferior thyroid artery is a branch of the thyrocervical trunk (a branch of the subclavian artery).

Which third thyroid artery exists in 10% of people?

In a small percentage of the population, the thyroid ima artery—an unpaired branch directly from the aorta, brachio-cephalic trunk, or left common carotid artery—ascends anterior to the thyroid and supplies the isthmus. This third artery can be a source of serious bleeding in patients undergoing thyroid surgery.

Describe the venous drainage of the thyroid.

Superior and middle thyroid veins drain to the internal jugular vein; inferior thyroid vein to the brachiocephalic vein

What is the venous drainage of the thyroid?

The superior and middle thyroid veins drain to the internal jugular vein. The inferior thyroid vein drains to the brachiocephalic veins.

Which important nerve runs with the inferior thyroid artery?

The recurrent laryngeal nerve—it is important to avoid this nerve during surgery!

PARATHYROID GLANDS

How many parathyroid glands are normally present?

Four (i.e., left and right superior and inferior)

Where do the parathyroid arteries originate?

They are usually branches from the inferior thyroid arteries.

Loss of the parathyroid glands results in what abnormality?

Hypocalcemia, which can lead to tetany (muscle spasm)

LARYNX

What is the colloquial name for the larynx?

The "voice box"

At what vertebral level is the larynx located?

At the level of vertebrae C3–C6

What two structures does the larynx connect?

The inferior pharynx and the trachea

Laryngeal cartilages

What tissue comprises the laryngeal skeleton?

Cartilage

How many cartilages are there in the larynx?

Nine, three paired and three unpaired

Name the three unpaired cartilages.

1. The thyroid cartilage
2. The cricoid cartilage
3. The epiglottic cartilage

What is the anatomic term for the "Adam's apple?"

The laryngeal prominence (i.e., where the lamina of the thyroid cartilage meet in the median plane)

What is the derivation of the word *thyroid*?

Greek *thyros* ("shield")

Which cartilage is palpable inferior to the thyroid cartilage?

The cricoid cartilage

Describe the attachments of the epiglottis:

The front of the epiglottis attaches to?

The body of the hyoid bone

The stalk attaches to?

The thyroid cartilage

The sides attach to?

The aryepiglottic folds

The upper edge attaches to?

Nothing

Which ligaments anchor the cricoid cartilage?

Superiorly: The cricothyroid ligaments
Inferiorly: The cricotracheal ligaments

Identify the labeled structures on the following figure of the larynx:

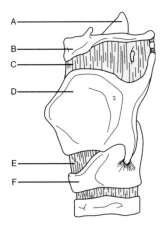

A = Epiglottic cartilage (i.e., epiglottis)
B = Body of the hyoid bone
C = Thyrohyoid membrane
D = Thyroid cartilage
E = Cricothyroid ligament
F = Cricoid cartilage

Which two structures pierce the thyrohyoid membrane on either side?	The internal laryngeal nerve and the superior laryngeal vessels
Name the three paired cartilages of the larynx.	1. Arytenoid cartilages 2. Corniculate cartilages 3. Cuneiform cartilages
Which of these pairs is most important in phonation, and how do they work?	The vocal cords attach to the pyramid-shaped arytenoid cartilages, which articulate with the cricoid cartilage. These articulations allow the arytenoid cartilages to slide, rotate, and tilt, thus modifying the position of (and the tension on) the vocal cords.
Where are the corniculate and cuneiform cartilages located?	Near the posterior aspects of the arytenoid cartilages

Laryngeal inlet

What are the aryepiglottic folds?	Folds of tissue containing cuneiform cartilage that form the posterior border of the inlet to the larynx (the epiglottis forms the anterior border)
What are the piriform recesses?	Named for their pear-like shape, the piriform recesses are recesses on each

side of the inlet to the larynx. They are separated from the inlet by the aryepiglottic folds.

What are the valleculae?

Two small (peanut-sized) depressions between the epiglottis and the anterior larynx; clinically significant because they are a frequent lodging place for food

Vocal cords

Vocal folds

What are the "true vocal cords?"

Vocal folds, consisting of the vocal ligaments and the conus elasticus

What is the conus elasticus?

An elastic membrane between the vocal ligaments and the cricoid cartilage (in other words, the vocal ligaments form the free edge of the conus elasticus)

Where do the vocal ligaments attach?

Between the vocal processes of the arytenoid cartilages and the posterior aspect of the thyroid cartilage

What is the sensory innervation of the larynx:

Above the vocal folds?

The internal branch of the superior laryngeal nerve, a branch of CN X (the vagus nerve)

Below the vocal folds?

The recurrent laryngeal branch of CN X

Vestibular folds

What are the "false vocal cords?"

Vestibular folds; these structures do not participate in sound production but appear similar to the true vocal cords

Where are the vestibular folds located in relation to the vocal folds?

They are superior to the vocal folds.

What comprises the vestibular folds?

Vestibular ligaments enclosed in a mucous membrane

Where do the vestibular ligaments attach?

Between the arytenoid cartilages and the point where the epiglottic cartilage meets the thyroid cartilage

The vestibular ligaments form the free edge of which membrane?

The quadrangular membrane, a membrane between the arytenoid cartilages and the epiglottic cartilage

Musculature of the larynx

Which two groups of muscles comprise the laryngeal musculature?

1. Extrinsic muscles (move the entire larynx)
2. Intrinsic muscles (move the small cartilages and vocal cords)

Extrinsic muscles

What are the three muscles of the inlet of the larynx and what do they do?

1. The transverse arytenoid muscle: Joins the arytenoid cartilages (this muscle is the only unpaired muscle of the larynx)
2. The oblique arytenoid muscles: Narrow the inlet
3. The thyroepiglottic muscles: Widen the inlet

Intrinsic muscles

Which muscles are the principal:

Adductors of the vocal cords?

The lateral cricoarytenoid muscles

Abductors of the vocal cords?

The posterior cricoarytenoid muscles

Tensors of the vocal cords?

The cricothyroid muscles

Relaxers of the vocal cords?

The thyroarytenoid muscles

Which nerve innervates all but one of the intrinsic laryngeal muscles?

The recurrent laryngeal nerve

Which muscle is the exception, and which nerve innervates it?

The cricothyroid muscle is innervated by the external branch of the superior laryngeal nerve.

What is the path of the recurrent laryngeal nerves in the neck?

They travel down as part of CN X (the vagus nerve) within the carotid sheath, and then branch around the great vessels (i.e., the brachiocephalic artery on the right and the arch of the aorta on the left) to travel up the tracheoesophageal groove. They then pierce the cricothyroid muscle and enter the larynx.

POWER REVIEW

MUSCLES AND FASCIAE OF THE NECK

What are the two insertions of the sternocleidomastoid muscle?	The sternum and the clavicle
What is the most superficial muscle of the neck, and which nerve innervates it?	The platysma, innervated by CN VII (the facial nerve)
How are the anterior muscles of the neck classified?	As suprahyoid or infrahyoid

Which four muscles comprise the suprahyoid group?

1. Mylohyoid
2. Geniohyoid
3. Stylohyoid
4. Digastric

What is the origin, insertion, and innervation of the digastric muscle?

Origin: The temporal bone, deep to the mastoid process
Insertion: Mandible
Innervation: CN VII (posterior belly); CN V (anterior belly)

Which four muscles comprise the infrahyoid group?

1. Omohyoid
2. Thyrohyoid
3. Sternohyoid
4. Sternothyroid

What is the other name for this group of muscles?	The "strap" muscles
Which nerve innervates all of the strap muscles but one?	The ansa cervicalis

Which muscle is the exception, and which nerve innervates this muscle?

The thyrohyoid, which is innervated by spinal nerve C1 via CN VII (the hypoglossal nerve)

What four deep muscles form the floor of the posterior triangle?

1. Splenius capitis
2. Levator scapulae
3. Posterior scalenus
4. Middle scalenus

What travels in the carotid sheath?

The common carotid artery, the internal jugular vein, and CN X (the vagus nerve)

What travels just outside the carotid sheath?

The sympathetic trunk

Disruption of this trunk causes what triad of symptoms?

Ptosis, miosis, and anhydrosis (i.e., Horner's syndrome)

What two structures pass between the anterior and middle scalenus muscles?

The brachial plexus and the subclavian artery

Vasculature of the neck

What are the branches of the internal carotid artery in the neck?

There are none; the ophthalmic artery (in the cranium) is the first branch

What are the six major branches of the external carotid artery, from proximal to distal?

1. Superior thyroid artery
2. Ascending pharyngeal artery
3. Lingual artery
4. Facial artery
5. Occipital artery
6. Posterior auricular artery

How does the external carotid artery terminate?

By dividing into the maxillary and superficial temporal arteries

What are the three major branches of the thyro-cervical trunk?

1. Inferior thyroid artery
2. Transverse cervical artery
3. Suprascapular artery

INNERVATION OF THE NECK

What are the four major sensory branches of the cervical plexus?	1. Lesser occipital nerve 2. Great auricular nerve 3. Transverse cervical nerve 4. Supraclavicular nerve
What is the origin and course of the phrenic nerve?	After originating from spinal nerves C3–C5, the phrenic nerve travels on the anterior surface of the anterior scalenus muscle to enter the thoracic inlet.

VISCERA OF THE NECK

What is the arterial supply and venous drainage of the thyroid gland?	**Arterial supply:** Inferior and superior thyroid arteries (branches of the thyrocervical trunk and the external carotid arteries, respectively) **Venous drainage:** Superior and middle thyroid veins drain to the internal jugular vein; inferior thyroid vein drains to the brachiocephalic vein
Where does the arterial supply of the parathyroid glands originate?	Usually from the inferior thyroid arteries
What is the anatomic term for the "true" vocal cords and where do they attach?	The vocal folds stretch between the vocal processes of the arytenoid cartilages and the posterior thyroid cartilage.
What is the anatomic term for the "false" vocal cords?	Vestibular folds
Which nerve innervates all but one of the intrinsic laryngeal muscles?	The recurrent laryngeal nerve
What muscle is the exception, and which nerve innervates it?	The cricothyroid muscle, which is innervated by the superior laryngeal nerve

6   The Back

VERTEBRAE

How many vertebrae are in each region of the back?	**Cervical region:** 7 **Thoracic region:** 12 **Lumbar region:** 5 **Sacral region:** 5 (fused) **Coccygeal region:** 4 (fused)
Which two curves of the vertebral column are:	
Concave anteriorly?	The thoracic and sacral curves
Concave posteriorly?	The cervical and lumbar curves
Which two curves are known as the primary curvatures?	The thoracic and sacral curves
Why?	The thoracic and sacral curves develop in the fetal period. (Recall that the fetus is "C"-shaped.)
Which two curves are known as the secondary curvatures?	The cervical and lumbar curves
When does the cervical curvature form?	When the infant holds its head erect (at 3–4 months)
When does the lumbar curvature form?	When the infant begins to walk (at the end of the first year)

Identify the labeled structures on the following views of a "typical" vertebra:

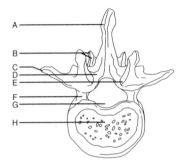

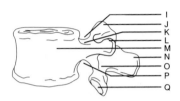

A = Spinous process
B = Inferior articular process and facet
C = Transverse process
D = Lamina
E = Superior articular facet
F = Pedicle
G = Vertebral foramen
H = Vertebral body
I = Superior vertebral notch
J = Superior articular process
K = Pedicle
L = Transverse process
M = Vertebral body
N = Spinous process
O = Lamina
P = Inferior vertebral notch
Q = Inferior articular facet

Which structures form the vertebral arch?

The pedicles (laterally) and the fused lamina (posteriorly)

What is the function of the vertebral arch?

Protection of the spinal cord, nerve roots, and meninges

How many processes arise from the vertebral arch of a typical vertebra?

Seven (two transverse processes, one spinous process, and four articular processes)

What does the vertebral arch form, along with the vertebral body?

The vertebral foramen

What is the vertebral canal?

The canal formed by the articulated vertebral foramina of successive vertebrae and the intervening intervertebral disks

Which structures pass through the intervertebral foramen?

The spinal nerve and artery and the intervertebral vein

Which cervical vertebra does not have a spinous process?

Vertebra C1

Describe the appearance of the spinous processes of vertebrae C2–C6.

Short and bifid

Why is vertebra C7 sometimes called the vertebra prominens?

It has the longest spinous process

What are the unique characteristics of vertebra C1 (the atlas)?

Vertebra C1 lacks a body and a spinous process. The anterior and posterior arches form the "top" and "bottom" of the vertebra, connecting the lateral masses that form its sides.

What is the odontoid process (dens)?

The portion of vertebra C2 (the axis) that projects superiorly and acts as a pivot point for the atlas

What is the colloquial name for an odontoid fracture?

A hangman's fracture

What structure is unique to the cervical vertebrae?

The paired transverse foramina

Which structures pass through the transverse foramina?

The vertebral artery, as well as the vertebral vein and autonomic nerves

Which cervical vertebra does not transmit vertebral arteries through its transverse foramina?

Vertebra C7

Which structures are unique to the thoracic vertebrae?

The costal facets (where the thoracic vertebrae articulate with the ribs)

What is unique about the lumbar vertebrae?

1. They have the largest bodies and pedicles.

2. A mamillary process is located on the posterior surface of each superior articular process. (Note: vertebra T12 also has mamillary processes.)

What is the sacral promontory?

The anterior edge of vertebra S1 (forms the posterior boundary of the true pelvis)

JOINTS AND LIGAMENTS

CRANIOVERTEBRAL JOINTS

What is the joint between the skull and vertebra C1 (the atlas) called?

The atlanto-occipital joint

Which motion occurs at the atlanto-occipital joint?

Flexion and extension of the head (nodding)

Which ligaments attach vertebra C1 to the skull?

The anterior and posterior atlanto-occipital membranes

What is the joint between the atlas and the axis called?

The atlanto-axial joint

What does the atlanto-axial joint consist of?

Two facet (plane) joints and one pivot joint between the dens and the anterior arch of the atlas

Which motion occurs at the atlanto-axial joint?

Rotation of the head from side to side Remember the motions of the atlanto-occipital and atlanto-axial joints by remembering "First yes (atlas), then no (axis)."

Where does the transverse ligament attach?

It runs between the tubercles on the lateral masses of vertebra C1, arching over the dens of vertebra C2

What is its purpose?

The transverse ligament holds the dens against the anterior arch of vertebra C1.

What are the points of insertion of the cruciform ligament?

Horizontally: The lateral masses of vertebra C1
Superiorly: The occipital bone
Inferiorly: The body of vertebra C2

Where do the alar ligaments attach?

They run from the sides of the dens to the lateral margins of the foramen magnum

Which movement is prevented by the alar ligaments?

The alar ligaments check the rotation and side-to-side movement of the head.

JOINTS OF THE VERTEBRAL BODIES

Where are the intervertebral disks located?	Between the bodies of adjacent vertebrae
At what level is the most superior intervertebral disk found?	Between vertebrae C2 and C3 (there is no intervertebral disk between the atlas and the axis)
At what level is the most inferior intervertebral disk found?	Between vertebrae L5 and S1
What is the external covering of the intervertebral disk called?	The anulus fibrosus
What is the internal matrix of the intervertebral disk called?	The nucleus pulposus
Which two ligaments play the most important role in stabilizing the vertebral bodies?	The anterior and posterior longitudinal ligaments
Where does the anterior longitudinal ligament run?	Along the anterior aspect of the vertebral bodies and intervertebral disks, from the occipital bone to the sacrum
Where does the posterior longitudinal ligament run?	Along the posterior aspect of the vertebral bodies, within the vertebral canal
Which membrane is the posterior longitudinal ligament continuous with?	The tectorial membrane
What does the tectorial membrane cover?	The surface of the dens and the apical, alar, and cruciform ligaments

INTERVERTEBRAL JOINTS

Which two structures form the facet joints?	The superior and inferior articular processes of adjacent vertebrae
What is the function of the facet joints?	The facet joints allow flexion, extension, and rotation of the spine. They also contribute to the spine's ability to bear

weight and prevent anterior movement of the superior vertebra onto the inferior one.

Which ligament connects the lamina of adjacent vertebrae?

The ligamentum flavum

What does the ligamentum flavum do?

It contributes to the posterior boundaries of the intervertebral foramina and helps to straighten the vertebral column after flexion.

Which ligaments connect the spinous processes?

The interspinous and supraspinous ligaments

What is the ligamentum nuchae?

A median fibrous septum between the posterior neck muscles; it is attached to the atlas and the cervical spinous processes and is the upward extension of the supraspinous ligament

Which ligaments connect the transverse processes?

The intertransverse ligaments (these are most substantial in the lumbar region)

Identify the labeled ligaments and associated structures of the vertebral column on the following views:

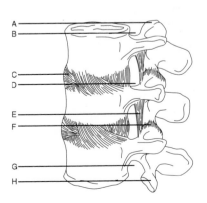

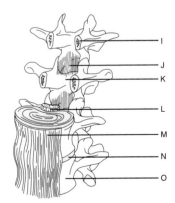

A = Superior articular process
B = Superior vertebral notch
C = Intervertebral disk
D = Intervertebral foramen
E = Ligamentum flavum
F = Articular capsule of the facet joint
G = Inferior vertebral notch
H = Inferior articular process
I = Pedicle
J = Ligamentum flavum
K = Lamina
L = Posterior longitudinal ligament
M = Anterior longitudinal ligament
N = Intervertebral disk
O = Vertebral body

When performing a lumbar puncture (spinal tap), which ligaments are pierced?

A lumbar puncture is usually performed at the level of vertebrae L3–L4 or L4–L5, in the midline between the iliac crests. After piercing the skin and superficial fascia, the needle passes through the supraspinous ligament, the interspinous ligament, and the ligamentum flavum before piercing the dura mater and the arachnoid mater to reach the cerebrospinal fluid (CSF).

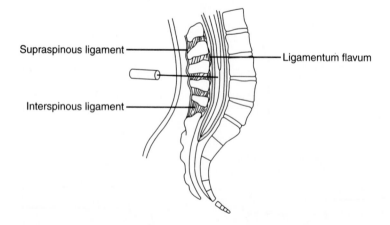

Supraspinous ligament

Ligamentum flavum

Interspinous ligament

MUSCLES

**What are the three ana-
tomic classifications used to
categorize the muscles of
the back?**

Superficial, intermediate, or deep

**What are the two functional
classifications used to
categorize the muscles of
the back?**

Extrinsic or intrinsic

**How do these groups
overlap?**

Extrinsic = superficial and intermediate
Intrinsic = deep

**Identify the superficial and
intermediate muscles of the
back and the related
structures on the following
figure:**

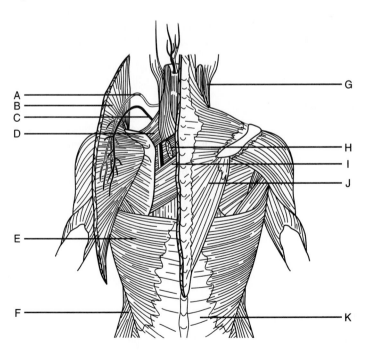

A = Accessory nerve
B = Trapezius muscle (reflected)
C = Transverse cervical artery
  (superficial branch)
D = Levator scapulae muscle
E = Latissimus dorsi muscle
F = External abdominal oblique muscle
G = Sternocleidomastoid muscle
H = Rhomboid minor muscle (cut)
I = Rhomboid major muscle
J = Trapezius muscle
K = Thoracolumbar fascia

SUPERFICIAL BACK MUSCLES

Latissimus dorsi muscle

Origin?

The spinous processes of vertebrae T7–T12, the thoracolumbar fascia (to vertebrae L1–L5), ribs 9–12, the upper sacral vertebrae, and the iliac crest

Insertion?

The floor of the bicipital groove of the humerus

Innervation?

The thoracodorsal nerve (from the brachial plexus; receives branches from the C6, C7, and C8 ventral rami)

Action?

Adducts, extends, and rotates the humerus medially at the shoulder joint

Trapezius muscle

Origin?

The external occipital protuberance, the superior nuchal line, the ligamentum nuchae, and the spinous processes of vertebrae C7–T12

Insertion?

The spine of the scapula, the acromion, and the lateral third of the clavicle

Innervation?

The spinal accessory nerve (CN XI) and branches of spinal nerves C3 and C4

Action?

Adducts, rotates, elevates, and depresses the scapula

Levator scapulae muscle

Origin?

The transverse processes of vertebrae C1–C4

Insertion?   The medial border of the scapula
                                        opposite the supraspinous fossa

Innervation?   The dorsal scapular nerve (from the
                                        brachial plexus; receives branches from
                                        the C5 ventral ramus)

Action?   Elevates the scapula

Rhomboid minor muscle

Origin?   The spinous processes of vertebra C7–T1

Insertion?   The root of the spine of the scapula

Innervation?   The dorsal scapular nerve

Action?   Adducts the scapula

Rhomboid major muscle

Origin?   The spinous processes of vertebrae
                                        T2–T5

Insertion?   The medial border of the scapula

Innervation?   The dorsal scapular nerve

Action?   Adducts the scapula

INTERMEDIATE BACK MUSCLES

Serratus posterior superior muscle

Origin?   The ligamentum nuchae, the
                                        supraspinous ligament, and the spinous
                                        processes of vertebrae C7–T3

Insertion?   The upper border of ribs 2–5

Innervation?   Intercostal nerves T1–T4 (i.e., the
                                        T1–T4 ventral primary rami)

Action?   Elevates the ribs

Serratus posterior inferior muscle

Origin?   The supraspinous ligament and the
                                        spinous processes of vertebrae
                                        T11–L3

Insertion?	The lower border of ribs 9–12
Innervation?	Intercostal nerves T9–T12 (i.e., the T9–T12 ventral primary rami)
Action?	Depresses the ribs

DEEP BACK MUSCLES

Name the three layers of deep muscles within the back.	1. Spinotransverse group (superficial) 2. Sacrospinalis group (intermediate) 3. Transversospinalis group (deep)
Which muscles comprise each group?	1. **Spinotransverse group:** Splenius capitis and splenius cervicis muscles 2. **Sacrospinalis group:** Erector spinae (formed by the iliocostalis, longissimus, and spinalis muscles) 3. **Transversospinalis group:** Semispinalis, multifidus, and rotatores muscles

Splenius capitis muscle

Origin?	The inferior half of the ligamentum nuchae and the spinous processes of vertebrae C7 and T1–T3
Insertions?	**Temporal bone:** At the mastoid process **Occipital bone:** Along the lateral third of the superior nuchal line
Innervation?	The dorsal rami of the inferior cervical nerves
Actions?	**Unilaterally:** Ipsilateral lateral flexion and rotation of the head and neck **Bilaterally:** Extension of the head and neck

Splenius cervicis muscle

Origin?	The spinous processes of vertebrae T3–T6
Insertion?	The transverse processes of vertebrae C1–C4

Innervation?

The dorsal rami of the inferior cervical nerves (the same as the splenius capitis muscle)

Actions?

Unilaterally: Ipsilateral lateral flexion and rotation of the head and neck
Bilaterally: Extension of the head and neck

Erector spinae

Is the erector spinae palpable?

Yes. The three vertical columns (i.e., the iliocostalis, longissimus, and spinalis muscles) form the prominent bulge that is palpable along each side of the vertebral column.

How are the iliocostalis, longissimus, and spinalis muscles subdivided?

Into three parts each, according to the superior attachments (e.g., spinalis thoracis, spinalis cervicis, spinalis capitis)

Which fascial compartment encloses the erector spinae?

The erector spinae lies between the posterior and anterior layers of the thoracolumbar fascia.

What is the common origin of the erector spinae?

Most of the divisions of the columns attach through a broad tendon to:
1. The posterior part of the iliac crest
2. The posterior part of the sacrum
3. The sacroiliac ligaments
4. The sacral and lumbar spinous processes

What is the insertion for the:

Iliocostalis muscle?

The ribs and cervical transverse processes

Longissimus muscle?

The ribs, transverse processes, and mastoid process

Spinalis muscle?

The spinous processes (note that the spinalis also arises from the spinous processes)

What is the action of the erector spinae:

Unilaterally?

Lateral flexion of the head or vertebral column

Bilaterally?

1. Extension of the vertebral column and the head
2. Control of movement during flexion

Semispinalis muscle

The semispinalis muscle is the largest muscle in which region?

The posterior neck

How many vertebral segments do fibers of this muscle span?

5–6

What are the three divisions of the semispinalis muscle?

The semispinalis thoracis, the semispinalis cervicis, and the semispinalis capitis

What is the origin of the:

Semispinalis thoracis?

The thoracic vertebrae

Semispinalis cervicis?

The cervical vertebrae

Semispinalis capitis?

The occipital bone between the inferior nuchal lines

What is the insertion of the:

Semispinalis thoracis?

The spinous processes of vertebrae C6–T4

Semispinalis cervicis?

The spinous processes of vertebrae C2–C5

Semispinalis capitis?

The planum nuchale (occipital bone)

Which nerves innervate the semispinalis muscle?

The dorsal rami of the spinal nerves

What is the bilateral action of the semispinalis muscle?

Extension of the head and upper vertebral column

What is the unilateral action of the semispinalis muscle?

Contralateral rotation of the head

Multifidus muscle

In which region is the multi-fidus muscle most prominent?

The lumbar region

Describe the origin and insertion of the multifidus muscle.

The multifidus muscle runs superomedially from the vertebral arches to the spinous processes, covering the laminae and spanning 3–4 vertebrae.

Innervation?

The dorsal rami of the spinal nerves

Actions?

Unilateral action: Ipsilateral flexion and contralateral rotation of the vertebral column
Bilateral action: Extension and stabilization of the spine

Rotatores muscles

Describe the origin and insertion of the rotatores.

The rotatores arise from the transverse process of one vertebrae and insert on the spinous process of the next (i.e., superior) vertebra.

Innervation?

The dorsal rami of the spinal nerves

Action?

Contralateral rotation and stabilization of the vertebral column

SUBOCCIPITAL REGION

What are the four major muscles of the suboccipital region?

1. Rectus capitis posterior major
2. Rectus capitis posterior minor
3. Obliquus capitis superior
4. Obliquus capitis inferior

What is the rectus capitis posterior major muscle's:

   Origin?

The spinous process of vertebra C2 (the axis)

   Insertion?

The lateral portion of the inferior nuchal line

What is the rectus capitis posterior minor muscle's:

   Origin?

The posterior tubercle of vertebra C2 (the axis)

Insertion?	The medial part of the inferior nuchal line

What is the obliquus capitis superior muscle's:

Origin?	The transverse process of vertebra C1 (the atlas)
Insertion?	The occipital bone, above the inferior nuchal line

What is the obliquus capitis inferior muscle's:

Origin?	The spinous process of vertebra C2 (the axis)
Insertion?	The transverse process of vertebra C1 (the atlas)

Which nerve innervates all of the suboccipital muscles?	The suboccipital nerve
Where does the suboccipital nerve originate?	The suboccipital nerve originates from the dorsal ramus of vertebra C1 (the atlas) and emerges between the vertebral artery (above) and the posterior arch of the atlas (below).
What actions do the suboccipital muscles perform as a unit?	1. Extension of the head (all four muscles) 2. Rotation of the head (all muscles except for the obliquus capitis superior) 3. Flexion of the head laterally (all muscles, except for the obliquus capitis inferior)

POWER REVIEW

What is the first easily palpated vertebra?	C7 (the long spinous process is palpable at the base of the neck)
Which feature is unique to the cervical vertebrae?	The paired transverse foramina

Do all of the transverse foramina transmit a vertebral artery?

No; only small accessory vertebral veins pass through the transverse foramina of vertebra C7.

What is the odontoid process (dens)?

The part of vertebra C2 (the axis) that projects superiorly from the vertebral body and articulates with vertebra C1 (the atlas)

Which action occurs at the:

    Atlanto-occipital joint?

Flexion and extension of the head (nodding)

    Atlanto-axial joint?

Lateral movement of the head (i.e., from side to side)

What does the cruciform ligament connect?

The vertical part connects vertebra C2 (the axis) to the foramen magnum. The horizontal portion spans vertebra C2 across the dens.

Where is the most superior intervertebral disk? The most inferior?

Most superior: Between vertebrae C2 and C3
Most inferior: Between vertebrae L5 and S1

What is the name of the external covering of the intervertebral disk? The soft internal part?

The anulus fibrosis (fibrous cartilage) and the nucleus pulposus (elastic cartilage), respectively

What are the five superficial muscles of the back?

1. Trapezius
2. Latissimus dorsi
3. Rhomboid major
4. Rhomboid minor
5. Levator scapulae (often considered with the upper limb)

Which two muscles comprise the spinotransverse (superficial) group of the deep back muscles?

The splenius capitis and splenius cervicis muscles

What comprises the sacrospinalis (intermediate) group of the deep back muscles?

The erector spinae (three parts)

Which three muscle groups comprise the deep group of the deep back muscles?

1. The semispinalis muscles (i.e., the semispinalis capitis, the semispinalis cervicis, and the semispinalis thoracis muscles)
2. The multifidus muscle
3. The rotatores muscles

7

The Upper Extremity

BONES

What are the five regions of the upper limb, and which bones are found in each region?

1. Pectoral girdle: Clavicle and scapula
2. Arm (brachium): Humerus
3. Forearm (antebrachium): Ulna and radius
4. Wrist (carpus): Carpal bones
5. Hand (manus): Metacarpal bones and phalanges

Identify the labeled bones on the following figure of the upper extremity:

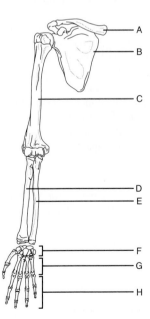

A = Clavicle
B = Scapula
C = Humerus
D = Radius
E = Ulna
F = Carpal bones
G = Metacarpal bones
H = Phalanges

PECTORAL GIRDLE

What is the function of the pectoral girdle?

The pectoral girdle connects the upper limb to the axial skeleton. (The axial skeleton consists of the skull, the vertebral column, the ribs and their cartilages, and the sternum.)

Which bone is the first to begin ossification during fetal development, but the last to complete it?

The clavicle (ossification begins at 7 weeks' gestation and ends 21 years after birth)

What are the two articulations of the clavicle?

The medial end of the clavicle articulates with the manubrium of the sternum at the sternoclavicular joint. The lateral end of the clavicle articulates with the acromion of the scapula at the acromioclavicular joint.

Describe the surface anatomy of the clavicle.

1. The jugular (suprasternal) notch lies between the medial elevations of the clavicles.
2. The medial two thirds of the clavicle is convex anteriorly; the large vessels and nerves that supply the upper limb pass posterior to the bone in this region.
3. The acromioclavicular joint can be palpated 2–3 centimeters medial to the acromion (i.e., the lateral extension of the spine of the scapula that forms the palpable "point" of the shoulder).

Where is the weakest point of the clavicle?

The point where the medial two thirds of the clavicle meet the lateral third

What is the name of the anterior surface of the scapula?

The subscapular fossa

Describe the coracoid process.

This bony process arises from the superior border of the scapula; it is often described as having a "bird's beak" appearance.

The coracoid process provides for which three muscular attachments?

1. The origin of the coracobrachialis muscle
2. The origin of the short head of the biceps brachii muscle
3. The insertion of the pectoralis minor muscle

Which bony landmark lies midway along the superior border of the scapula?

The suprascapular notch

Which ligament runs across the suprascapular notch?

The superior transverse scapular ligament

Which two structures traverse the superior transverse scapular ligament?

The suprascapular artery runs over the ligament, and the suprascapular nerve runs under the ligament. Think, "The Army goes over the bridge and the Navy goes under it."

Which structure divides the posterior surface of the scapula into two fossae?

The spine of the scapula divides the posterior surface of the scapula into the supraspinous fossa and the infraspinous fossa.

Which of these fossae is the largest?

The infraspinous fossa (i.e., the one below the spine of the scapula)

Which muscle originates from the:

Spine of the scapula?

The deltoid muscle

Supraspinous fossa?

The supraspinatus muscle

Infraspinous fossa?

The infraspinatus muscle

What are the articulations of the scapula?

The spine of the scapula continues laterally as the acromion, which articulates anteriorly with the clavicle to form the acromioclavicular joint. The lateral surface of the scapula forms the glenoid fossa, which articulates with the head of the humerus to form the glenohumeral joint.

What structure deepens the glenoid fossa?

The glenoid labrum, a fibrocartilaginous lip that extends over the glenoid fossa, thereby deepening it

Which muscles originate just above and below the glenoid fossa?

The long head of the biceps brachii and the long head of the triceps brachii, respectively

Identify the labeled structures on the following diagram of the scapula (posterior view):

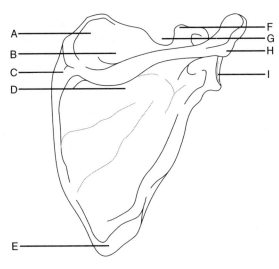

A = Superior angle
B = Supraspinous fossa
C = Root of the spine of the scapula
D = Infraspinous fossa
E = Inferior angle
F = Coracoid process
G = Suprascapular notch
H = Acromion
I = Glenoid cavity

Describe the surface anatomy of the scapula as it relates to the vertebral column.

1. The root of the spine of the scapula (i.e., the medial end) is opposite the spinous process of vertebra T3.
2. The superior angle of the scapula lies at the level of vertebra T2.

3. The inferior angle of the scapula lies at the level of vertebra T7.

ARM (BRACHIUM)

Identify the labeled structures on the following figure of the humerus (anterior view):

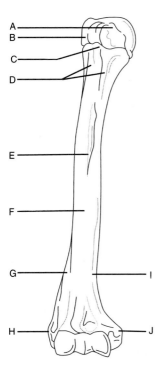

A = Lesser tubercle
B = Greater tubercle
C = Intertubercular groove
D = Surgical neck
E = Deltoid tuberosity
F = Humeral shaft (body)
G = Lateral supracondylar ridge
H = Lateral epicondyle
I = Medial supracondylar ridge
J = Medial epicondyle

Which three muscles insert on the greater tubercle of the humerus?

The supraspinatus, infraspinatus, and teres minor muscles (i.e., all the rotator cuff muscles except for the subscapularis muscle)

Which muscle inserts on the lesser tubercle of the humerus?

The subscapularis muscle (i.e., the fourth rotator cuff muscle)

Which three muscles insert on the intertubercular groove?

1. **Pectoralis major:** Lateral lip of the groove
2. **Teres major:** Medial lip of the groove
3. **Latissimus dorsi:** Floor of the groove
   Remember, three "major" muscles insert on the intertubercular groove: the pectoralis major, the teres major, and the largest muscle of the back, the latissimus dorsi.

Where is the anatomical neck of the humerus?

Distal to the head of the humerus and proximal to the greater and lesser tubercles

Where is the surgical neck of the humerus?

Distal to the greater and lesser tubercles, where the humeral shaft begins

Where is the radial (spiral) groove and why is it clinically important?

The radial groove runs obliquely along the posterior humerus and houses the radial nerve. Therefore, fractures affecting the shaft of the humerus may lead to radial nerve damage.

Name the two articular surfaces of the distal humerus.

1. The **capitulum** (i.e., the lateral articular surface) articulates with the head of the radius.
2. The **trochlea** (i.e., the medial articular surface) articulates with the trochlear notch of the ulna.

Where are the coronoid fossa and the olecranon fossa?

The coronoid fossa is superior to the trochlea on the distal end of the humerus anteriorly. The olecranon fossa lies in the same position posteriorly.

What structures do these fossae accommodate?

The coronoid process and the olecranon of the ulna, respectively

FOREARM (ANTEBRACHIUM)

Identify the labeled structures on the following figure of the bones of the forearm (anterior view):

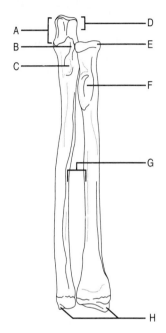

A = Trochlear notch
B = Coronoid process
C = Ulnar tuberosity
D = Olecranon
E = Head of the radius
F = Radial tuberosity
G = Interosseus borders
H = Styloid processes

Which styloid process is more distal, the radial styloid process or the ulnar styloid process?

The radial styloid process is normally approximately 1 centimeter distal to the ulnar styloid process.

Which muscle inserts on the styloid process of the radius?

The brachioradialis muscle

Which muscle inserts on the:

Radial tuberosity?

The biceps brachii muscle

Ulnar tuberosity?

The brachialis muscle

**Describe the articulations
of the:**

    Proximal radius

The capitulum of the humerus and the
radial notch of the ulna

    Distal radius

The proximal row of carpal bones (except
for the pisiform bone)

**Where are the heads of the
radius and ulna located?**

The head of the radius is located at the
proximal end of the bone, articulating
with the capitulum of the humerus and
the radial notch of the ulna. The head of
the ulna is at the distal end of the bone,
articulating with the articular disk of the
radioulnar joint. To remember the
location of the radial and ulnar heads,
think **RPUD: R**adial **P**roximal, **U**lna
Distal.

**What attaches the lateral
aspect of the shaft of the
ulna and the medial aspect
of the shaft of the radius?**

The interosseous membrane

**What other function does
the interosseous membrane
serve?**

It is the attachment site for several of the
deep forearm muscles.

WRIST (CARPUS)

**Identify the labeled bones
on the following figure of
the wrist (posterior view):**

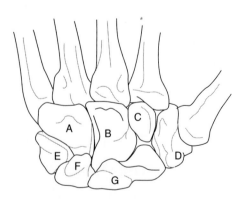

A = Hamate
B = Capitate
C = Trapezoid
D = Trapezium
E = Triquetrum
F = Lunate
G = Scaphoid
Note that the pisiform, which lies just anterior to the triquetrum, is not visible in a posterior view.

What is a mnemonic to remember the carpal bones?

From lateral to medial, proximal row followed by distal row:
Some **L**overs **T**ry **P**ositions **T**hat **T**hey **C**an't **H**andle
Scaphoid
Lunate
Triquetrum
Pisiform
Trapezium
Trapezoid
Capitate
Hamate
To remember that the trapezium comes before the trapezoid, think "trapezium with thumb."

Which of the four carpal bones in the proximal row does not articulate with the radius and articular disk?

The pisiform

Which of the four carpal bones in the proximal row articulates with the ulna?

None of the carpal bones articulates with the ulna.

Which carpal bone is commonly fractured when one falls on the palm with the hand abducted?

The scaphoid

Which carpal bone is the most commonly fractured?

The scaphoid

HAND (MANUS)

**Identify the labeled bones
on the following figure of
the wrist and hand
(anterior view):**

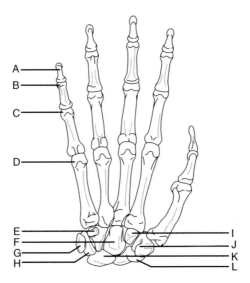

A = Distal phalanx
B = Head of the middle phalanx
C = Head of the proximal phalanx
D = Head of the fifth metacarpal bone
E = Hook of the hamate
F = Capitate
G = Pisiform
H = Triquetrum
I = Trapezoid
J = Tubercle of the trapezium
K = Lunate
L = Tubercle of the scaphoid

**Name the three parts of a
metacarpal bone and the
location of each.**

The base is proximal, the head is distal,
and the shaft (body) is in the middle.

How many phalanges are in:

Each finger?

Three (proximal, middle, and distal)

The thumb?

Two (proximal, distal)

PECTORAL GIRDLE AND SHOULDER

PECTORAL MUSCLES

Name the four pectoral muscles.

1. Pectoralis major muscle
2. Pectoralis minor muscle
3. Serratus anterior muscle
4. Subclavius muscle

Pectoralis major muscle

Origin?

Clavicular head: The anterior surface of the medial clavicle

Sternocostal head: The anterior surface of the sternum and the superior six costal cartilages

Insertion?

The lateral lip of the intertubercular groove of the humerus

Innervation?

The medial and lateral pectoral nerves (branches of the brachial plexus receiving fibers from the C8 and T1 ventral rami and the C5, C6, and C7 ventral rami, respectively)

Action?

Adducts and medially rotates the humerus at the glenohumeral (shoulder) joint

The superior border of the pectoralis major muscle contributes to which anatomic triangle?

The deltopectoral triangle (the other sides are formed by the deltoid, one of the scapular muscles, and the clavicle)

What vascular structure lies in the deltopectoral triangle?

The cephalic vein

Pectoralis minor muscle

Origin?

Ribs 2–5

Insertion?

The coracoid process of the scapula

Innervation?

The medial pectoral nerve

Action?

Stabilizes the scapula by drawing it anteriorly

The pectoralis minor muscle divides which important axillary structure into three parts?	The axillary artery

Serratus anterior muscle

Origin?	The external surfaces of ribs 1–8 (the muscle is named for the saw-toothed appearance of its proximal attachments)
Insertion?	The medial border of the scapula
Innervation?	The long thoracic nerve (a branch of the brachial plexus receiving fibers from the C5, C6, and C7 ventral rami)
Action?	Holds the scapula against the thoracic wall
What happens when the long thoracic nerve is injured?	Denervation of the long thoracic nerve, leading to loss of serratus anterior function, characteristically results in a phenomenon known as "winging of the scapula"
What is the blood supply of the serratus anterior muscle?	The dorsal scapular artery (a branch of the subclavian artery)

Subclavius muscle

Origin?	Rib 1, at the costal cartilage
Insertion?	The inferior surface of the clavicle
Innervation?	The nerve to subclavius (a branch of the brachial plexus receiving fibers from the C5 and C6 ventral rami)
Action?	Stabilizes the clavicle

SCAPULAR MUSCLES

Which six muscles pass from the scapula to the humerus and act on the shoulder joint?	1. Deltoid muscle 2. Teres major muscle 3. Supraspinatus muscle 4. Infraspinatus muscle

5. Teres minor muscle
6. Subscapularis muscle

Which of these muscles comprise the "rotator cuff" muscles?

SITS
Supraspinatus
Infraspinatus
Teres minor
Subscapularis

Why are these four muscles known as the "rotator cuff" muscles?

Along with their corresponding tendons, these four muscles surround the glenohumeral joint, forming a musculotendinous "cuff" that protects and stabilizes the joint by holding the head of the humerus in the glenoid fossa.

Deltoid muscle

Origin?

The lateral third of the clavicle, the acromion of the scapula, and the spine of the scapula

Insertion?

The deltoid tuberosity of the humerus

Innervation?

The axillary nerve (a terminal branch of the brachial plexus receiving fibers from the C5 and C6 ventral rami)

Action?

Anterior part: Flexion and medial rotation of the humerus at the glenohumeral joint
Middle part: Abduction of the humerus at the glenohumeral joint
Posterior part: Extension and lateral rotation of the humerus at the glenohumeral joint

Teres major muscle

Origin?

The dorsal surface of the inferior angle of the scapula

Insertion?

The medial lip of the intertubercular groove of the humerus

Innervation?

The lower subscapular nerve (a branch of the brachial plexus receiving fibers from the C6 and C7 ventral rami)

Action?

Adduction and medial rotation of the humerus at the glenohumeral joint

What surface landmark is formed by the teres major muscle and the tendon of the latissimus dorsi muscle?

The posterior axillary fold

Supraspinatus muscle

Origin?

The supraspinous fossa of the scapula

Insertion?

The superior facet of the greater tubercle of the humerus

Innervation?

The suprascapular nerve (a branch of the brachial plexus receiving fibers from the C4, C5, and C6 ventral rami)

Action?

Assists the deltoid with abduction of the humerus at the glenohumeral joint; acts in concert with the other rotator cuff muscles to hold the head of the humerus in the glenoid fossa

Infraspinatus muscle

Origin?

The infraspinous fossa of the scapula

Insertion?

The middle facet on the greater tubercle of the humerus

Innervation?

The suprascapular nerve

Action?

Lateral rotation of the humerus at the glenohumeral joint; also assists in holding the head of the humerus in the glenoid fossa

Teres minor muscle

Origin?

The superior part of the lateral border of the scapula

Insertion?

The inferior facet on the greater tubercle of the humerus

Innervation?

The axillary nerve

Action?

Lateral rotation of the humerus at the glenohumeral joint; also assists in holding the head of the humerus in the glenoid fossa (note that the action of the teres minor muscle is identical to that of the infraspinatus muscle)

Subscapularis muscle

Origin?

The subscapular fossa on the anterior surface of the scapula

Insertion?

The lesser tubercle of the humerus

Innervation?

The upper and lower subscapular nerves (branches of the brachial plexus receiving fibers from the C5, C6, and C7 ventral rami)

Action?

Medial rotation and adduction of the humerus at the glenohumeral joint; also assists in holding the head of the humerus in the glenoid fossa

FASCIAE

The pectoral fascia is continuous with which structure inferiorly?

The fascia of the abdominal wall

Laterally, the pectoral fascia becomes what?

The axillary fascia

Which two muscles are enveloped by the clavipectoral fascia?

The subclavius and pectoralis minor muscles

What is the portion of the clavipectoral fascia between the first rib and the coracoid process of the scapula called?

The costocoracoid membrane

Which artery, vein, and nerve pierce the costocoracoid membrane?

The thoracoacromial artery (a branch of the axillary artery), the cephalic vein, and the lateral pectoral nerve

JOINTS AND LIGAMENTS OF THE PECTORAL GIRDLE

What are the three joints of the pectoral girdle?

1. Sternoclavicular joint
2. Acromioclavicular joint
3. Glenohumeral (shoulder) joint

Describe the sterno-clavicular joint.

The sternoclavicular joint is a synovial joint. The synovial membrane lines the articular capsule between the sternal end of the clavicle and the manubrium of the sternum.

Name the four ligaments of the sternoclavicular joint and describe their functions.

1. **Anterior sternoclavicular ligament:** Reinforces the capsule anteriorly
2. **Posterior sternoclavicular ligament:** Reinforces the capsule posteriorly
3. **Interclavicular ligament:** Stabilizes the medial ends of the clavicles
4. **Costoclavicular ligament:** Attaches the inferior surface of the medial end of the clavicle with the first rib and its costal cartilage

Which arteries supply the sternoclavicular joint?

The internal thoracic and suprascapular arteries (branches of the subclavian artery and the thyrocervical trunk, respectively)

Describe the acromio-clavicular joint.

The acromioclavicular joint is a synovial joint that joins the lateral end of the clavicle with the acromion of the scapula.

Name the two ligaments of the acromioclavicular joint.

1. Acromioclavicular ligament
2. Coracoclavicular ligament (formed by the conoid and trapezoid ligaments)

Which arteries supply the acromioclavicular joint?

The suprascapular and thoracoacromial arteries

Describe the glenohumeral joint.

The glenohumeral joint is a ball-and-socket joint between the glenoid fossa of the scapula and the head of the humerus.

Name the six ligaments of the glenohumeral joint.

1. Superior glenohumeral ligament
2. Middle glenohumeral ligament
3. Inferior glenohumeral ligament
4. Transverse humeral ligament
5. Coracohumeral ligament
6. Coracoacromial ligament

The glenohumeral joint allows for which types of movements?

Flexion and extension, abduction and adduction, medial and lateral rotation, and circumduction

Name the three nerves that innervate the glenohumeral joint.	1. Axillary nerve 2. Suprascapular nerve 3. Lateral pectoral nerve
Name the four arteries that supply the glenohumeral joint.	1. Anterior circumflex humeral artery 2. Posterior circumflex humeral artery 3. Suprascapular artery 4. Scapular circumflex artery
Name the three bursae of the glenohumeral joint.	1. Subacromial bursa 2. Subdeltoid bursa 3. Subscapular bursa
What is the function of these bursae?	To reduce friction between the rotator cuff and the coracoacromial arch during movement of the glenohumeral joint

Name the muscles involved in the following movements at the glenohumeral joint:

Adduction	Pectoralis major, latissimus dorsi, teres major, triceps, and subscapularis
Abduction	Deltoid and supraspinatus
Flexion	Pectoralis major, anterior part of the deltoid, coracobrachialis, and biceps brachii
Extension	Latissimus dorsi, posterior part of the deltoid, triceps, and teres major
Medial rotation	Subscapularis, pectoralis major, anterior part of the deltoid, latissimus dorsi, and teres major
Lateral rotation	Infraspinatus, teres minor, and posterior part of the deltoid

AXILLA

What is the axilla?	The pyramidal area at the junction of the upper extremity and the trunk (underarm)
What are the boundaries of the axilla:	
Medially?	Ribs 1–4, the intercostal muscles, and the serratus anterior muscle

Laterally?	The humerus (specifically, the floor of the intertubercular groove)
Anteriorly?	The pectoralis major and pectoralis minor muscles
Posteriorly?	The scapula and the subscapularis, teres major, and latissimus dorsi muscles
What forms the base of the axilla?	The axillary fascia and skin
The apex?	The interval between the clavicle, scapula, and first rib
What six structures are contained within the axilla?	1. Axillary artery 2. Axillary vein 3. Axillary lymph nodes 4. Branches of the brachial plexus 5. Biceps brachii muscle (the long and short heads) 6. Coracobrachialis muscle
What is the axillary sheath?	A continuation of the cervical fascia into the axilla that encloses the axillary artery, the axillary vein, and the brachial plexus

VASCULATURE

Axillary artery

Describe the origin and fate of the axillary artery.	The subclavian artery becomes the axillary artery at the lateral border of the first rib. The axillary artery becomes the brachial artery at the inferior border of the teres major muscle.
Delineate the three parts of the axillary artery, and name the branches from each. (Note that the first part has one branch, the second part two, and the third part three!)	
First part?	Extends from the lateral border of the first rib to the superior border of the pectoralis minor muscle, giving off the superior thoracic artery

Second part?

Extends deep to the pectoralis minor muscle, giving off the thoracoacromial artery and the lateral thoracic artery

Third part?

Extends from the inferior border of the pectoralis minor muscle to the inferior border of the teres major muscle, giving off the subscapular, anterior circumflex humeral, and posterior circumflex humeral arteries

What are the two branches of the subscapular artery?

1. Circumflex scapular artery
2. Thoracodorsal artery

Which two branches of the axillary artery anastomose with one another on the surgical neck of the humerus?

The anterior and posterior circumflex humeral arteries

Identify the labeled arteries on the following figure:

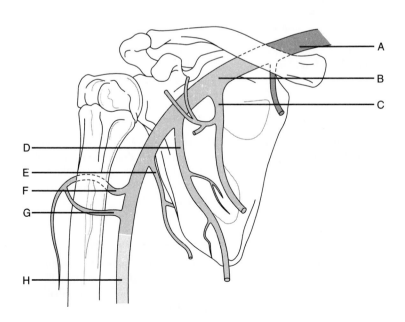

A = Subclavian artery
B = Axillary artery
C = Thoracoacromial artery
D = Lateral thoracic artery
E = Subscapular artery
F = Posterior circumflex humeral artery
G = Anterior circumflex humeral artery
H = Brachial artery
Note that the superior thoracic artery branches off the first part of the axillary posteriorly and therefore is not visible in this view.

Axillary vein

Describe the origin and fate of the axillary vein.

The axillary vein originates at the inferior border of the teres major muscle as a continuation of the basilic vein, and becomes the subclavian vein at the lateral border of the first rib. Note that the borders of the axillary vein parallel those of the axillary artery.

Within the axilla, what is the relationship between the axillary vein and artery?

The vein lies superficial to the artery.

Axillary lymph nodes

Name the five groups of axillary lymph nodes.

1. Central
2. Lateral
3. Subscapular (posterior)
4. Pectoral
5. Apical

What is the course of lymphatic drainage in the axillary region?

The lateral, pectoral, and subscapular (posterior) nodes drain into the central nodes, which in turn drain into the apical nodes. The apical nodes drain into the subclavian trunks.

BRACHIAL PLEXUS

What is the brachial plexus?

A large network of nerves that originates in the neck and extends into the axilla, giving rise to most of the nerves that supply the upper extremity

What are the five segmental branchings of the brachial plexus?	**R**on **T**aylor **D**rinks **C**old **B**eer **R**ami **T**runks **D**ivisions **C**ords **B**ranches (terminal)
The ventral primary rami of which spinal cord segments contribute to the brachial plexus?	C5, C6, C7, C8, and T1
The rami leading to the brachial plexus run between which two muscles?	The anterior and middle scalene muscles
Which two nerves branch off from the rami of the brachial plexus before the rami become trunks?	1. The dorsal scapular nerve: From the C5 rami, supplies the rhomboid minor and rhomboid major muscles 2. The long thoracic nerve: From the C5, C6, and C7 rami, supplies the serratus anterior muscle (remember scapular winging?)
Which rami contribute to which trunks?	**Superior trunk:** Formed from the rami of C5 and C6 **Middle trunk:** Continuation of the ramus of C7 **Inferior trunk:** Formed from the rami of C8 and T1
Which two nerves branch off the superior trunk of the brachial plexus?	1. The suprascapular nerve: From the C5 and C6 rami, supplies the supraspinatus and infraspinatus muscles and the glenohumeral joint, runs under the superior transverse scapular ligament (remember, the **N**erve is like the **N**avy!) 2. The nerve to subclavius: From the C5 ramus, supplies the subclavius muscle
What do the trunks divide into?	Each trunk splits into an anterior division and a posterior division.
How are the cords formed from the anterior and posterior divisions of the trunks?	**Lateral cord:** Formed from the anterior divisions of the superior and middle trunks **Posterior cord:** Formed from the posterior divisions of all three trunks

Medial cord: Continuation of the anterior division of the inferior trunk

The lateral cord gives rise to which branch?

The lateral pectoral nerve: From the C5, C6, and C7 rami, supplies the pectoralis major muscle and contributes to the innervation of the pectoralis minor muscle

What are the two terminal branches of the lateral cord?

1. The musculocutaneous nerve: From the C5, C6, and C7 rami, supplies the coracobrachialis, biceps brachii, and brachialis muscles
2. The lateral root of the median nerve

The medial cord gives rise to which three branches?

1. The medial pectoral nerve: From the C8 and T1 rami, supplies the pectoralis minor and pectoralis major muscles
2. The medial brachial cutaneous nerve: From the C8 and T1 rami, supplies the skin on the medial side of the arm
3. The medial antebrachial cutaneous nerve: From the C8 and T1 rami, supplies the skin on the medial side of the forearm

Where does the medial pectoral nerve lie in relation to the lateral pectoral nerve?

Lateral to it! The medial pectoral nerve is named "medial" because it arises from the medial cord of the brachial plexus, not because of its position relative to the lateral pectoral nerve. Similarly, the lateral pectoral nerve is designated "lateral" because it arises from the lateral cord; it lies medial to the medial pectoral nerve.

What are the two terminal branches of the medial cord?

1. The ulnar nerve
2. The medial root of the median nerve

Which terminal nerve branch of the brachial plexus receives contributions from both the medial and lateral cords?

The median nerve: From the C5, C6, C7, C8, and T1 rami

The posterior cord gives rise to which three branches?

1. The upper subscapular nerve: From the C5 and C6 rami, innervates the upper portion of the subscapularis muscle

2. The thoracodorsal nerve: From the C7 and C8 rami, innervates the latissimus dorsi muscle
3. The lower subscapular nerve: From the C5 and C6 rami, innervates the lower portion of the subscapularis muscle as well as the teres major muscle

What are the two terminal branches of the posterior cord?

1. The axillary nerve: From the C5 and C6 rami, innervates the deltoid and teres minor muscles, eventually becomes the lateral brachial cutaneous nerve, which supplies the skin on the lateral arm
2. The radial nerve: From the C5, C6, C7, C8, and T1 rami

Which nerve is the largest branch of the brachial plexus?

The radial nerve

Which muscles are innervated by the radial nerve?

The extensors of the upper limb

Identify the labeled nerves on the following diagram of the brachial plexus:

(Come on! You can do it!)

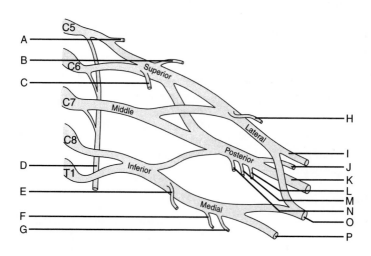

A = Dorsal scapular nerve
B = Suprascapular nerve
C = Nerve to subclavius
D = Long thoracic nerve
E = Medial pectoral nerve
F = Medial brachial cutaneous nerve
G = Medial antebrachial cutaneous nerve
H = Lateral pectoral nerve
I = Musculocutaneous nerve
J = Axillary nerve
K = Radial nerve
L = Lower subscapular nerve
M = Thoracodorsal nerve
N = Upper subscapular nerve
O = Median nerve
P = Ulnar nerve
Note: The brachial plexus is a key anatomic structure that will come up again and again during your anatomy course and beyond; you would be well served to memorize it so that you can reproduce a diagram of it on command.

ARM (BRACHIUM)

MUSCLES OF THE ARM

What divides the arm into anterior fascial (flexor) and posterior fascial (extensor) compartments?

The medial and lateral intermuscular septa and the humerus

Which muscles of the brachium lie in the:

Anterior fascial (flexor) compartment?

BBC
Brachialis muscle
Biceps brachii muscle
Coracobrachialis muscle

Posterior fascial (extensor) compartment?

Triceps brachii muscle

Brachialis muscle

Origin?

The distal half of the anterior surface of the humerus

Insertion?	The ulnar tuberosity and the coronoid process of the ulna
Innervation?	The musculocutaneous nerve
Action?	Main flexor of the forearm at the humeroulnar (elbow) joint

Biceps brachii muscle

Origin?	**Long head:** The supraglenoid tubercle of the scapula **Short head:** The coracoid process of scapula
Insertion?	The radial tuberosity and the fascia of the medial forearm via the bicipital aponeurosis
Innervation?	The musculocutaneous nerve
Action?	Supination of the radioulnar joints and flexion of the humeroulnar joint from a supine position
Describe the course of the tendon of the long head of the biceps brachii muscle.	The tendon of the long head of the biceps brachii muscle crosses the head of the humerus (enclosed in a fibrous capsule) and descends in the intertubercular groove to the radial tuberosity.

Coracobrachialis muscle

Origin?	The tip of the coracoid process of the scapula
Insertion?	The medial surface of the humerus, about halfway down
Innervation?	The musculocutaneous nerve
Action?	Assists in flexion and adduction of the arm at the glenohumeral joint

Triceps brachii muscle

Origin?	**Long head:** The infraglenoid tubercle of the scapula **Lateral head:** The posterior surface of the humerus, proximal (superior) to the radial groove

Medial head: The posterior surface of humerus, distal (inferior) to the radial groove

Insertion?

The olecranon of the ulna

Innervation?

The radial nerve (remember, the radial nerve innervates **all** of the extensors in the upper limb)

Action?

The triceps brachii is the chief extensor of the forearm at the humeroulnar joint; in addition, the long head of the triceps brachii steadies the head of the abducted humerus.

Which small muscle assists the triceps brachii muscle in extending the forearm?

The anconeus muscle, located in the forearm

VASCULATURE OF THE ARM

Arteries

Which artery represents the principal arterial supply to the upper limb?

The brachial artery

Describe the course of the brachial artery.

The brachial artery originates at the inferior border of the teres major muscle as a continuation of the axillary artery and courses distally in the "bicipital groove" anterior to the medial intermuscular septum before ending in the cubital fossa by dividing into the ulnar and radial arteries.

The superior ulnar collateral artery pierces which fascial membrane?

The medial intermuscular septum

The superior ulnar collateral artery travels with which nerve behind the medial epicondyle?

The ulnar nerve

Identify the labeled arteries on the following figure of the arm and forearm:

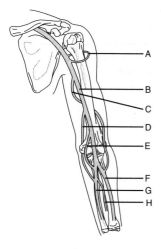

A = Anterior and posterior circumflex humeral arteries
B = Brachial artery
C = Deep brachial artery
D = Superior ulnar collateral artery
E = Inferior ulnar collateral artery
F = Radial artery
G = Ulnar artery
H = Anterior interosseus artery

Veins

Name the two main superficial veins of the arm.

The cephalic vein and the basilic vein

Where do the cephalic and basilic veins originate from?

The dorsal venous arch of the hand

Which superficial vein is more lateral?

The cephalic vein

Which superficial vein runs through the deltopectoral triangle?

The cephalic vein

Which vein forms a communication between the basilic and cephalic veins in the cubital fossa?

The median cubital vein

Identify the labeled superficial veins on the following figure of the arm and forearm:

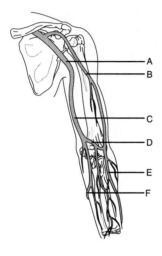

A = Axillary vein
B = Cephalic vein
C = Basilic vein
D = Median cubital vein
E = Cephalic vein
F = Basilic vein

INNERVATION OF THE ARM

Which four major nerves run through the arm?	1. Median nerve 2. Ulnar nerve 3. Radial nerve 4. Musculocutaneous nerve
Which of these do not branch in the arm?	The median and ulnar nerves
The median nerve lies in close proximity to which artery?	The brachial artery (the median nerve is located just anterior to it)
Describe the course of the radial nerve in the arm.	The radial nerve enters the arm posterior to the brachial artery and medial to the humerus. It passes inferolaterally along the humerus in the radial groove (accompanied by the deep brachial artery), and pierces the lateral intermuscular septum to run between the brachialis and brachioradialis muscles. At the lateral epicondyle of the humerus, the

radial nerve divides into deep and superficial branches.

Describe the course of the musculocutaneous nerve.

The musculocutaneous nerve pierces the coracobrachialis muscle and descends between the biceps brachii and brachialis muscles. Eventually, the musculo-cutaneous nerve becomes the lateral antebrachial cutaneous nerve. (Recall that the medial antebrachial cutaneous nerve is a branch from the medial cord of the brachial plexus.)

ELBOW REGION

CUBITAL FOSSA

What is the cubital fossa?

The hollow area on the anterior surface of the elbow

What are the boundaries of the cubital fossa:

Laterally?

The brachioradialis muscle (a superficial posterior muscle in the forearm)

Medially?

The pronator teres muscle (a superficial anterior muscle in the forearm)

Superiorly?

Imagine a line between the medial and lateral epicondyles of the humerus

What forms the floor of the cubital fossa?

The brachialis and supinator muscles

The roof?

The superficial fascia, skin, and bicipital aponeurosis

List four key structures found in the cubital fossa.

1. The brachial artery
2. The median nerve
3. The radial artery
4. The ulnar artery

Where does the median cubital vein lie in relation to the bicipital aponeurosis?

Superficial to it; the veins in the region of the cubital fossa are a favorite target for phlebotomists because of their superficial position

HUMEROULNAR (ELBOW) JOINT

Describe the humeroulnar joint.	The humeroulnar joint is a synovial joint in which the trochlea of the humerus articulates with the trochlear notch of the ulna, and the capitulum of the humerus articulates with the head of the radius.

Name the muscles involved in:

Flexion of the humero-ulnar joint?	The brachialis, biceps brachii, brachio-radialis, and pronator teres muscles
Extension of the humero-ulnar joint?	The triceps brachii and anconeus muscles
Name the three ligaments that reinforce the humero-ulnar joint.	1. Annular ligament 2. Radial collateral ligament 3. Ulnar collateral ligament
Describe the annular ligament.	The annular ligament travels around most of the head of the radius, preventing withdrawal of the head from its socket. The annular ligament then fuses with the radial collateral ligament.

FOREARM (ANTEBRACHIUM)

MUSCLES OF THE FOREARM

What are the two groups of forearm muscles?	1. The flexor-pronator (anterior forearm) group 2. The extensor-supinator (posterior forearm) group

Anterior forearm muscles

Classify the eight flexor-pronator muscles by action. Which flexor-pronator muscles:

Pronate the forearm and hand?	1. Pronator teres muscle 2. Pronator quadratus muscle
Flex the hand at the radiocarpal joint?	3. Flexor carpi radialis muscle 4. Flexor carpi ulnaris muscle 5. Palmaris longus muscle

Flex the interphalangeal joints of the digits?

6. Flexor digitorum superficialis muscle
7. Flexor digitorum profundus muscle
8. Flexor pollicis longus muscle

What is the only muscle that can flex the distal inter-phalangeal joints?

The flexor digitorum profundus muscle (this muscle also assists in flexion at the metacarpophalangeal and wrist joints)

Which muscle is most important in flexing the digits slowly?

The flexor digitorum superficialis muscle

Which muscle is most important in flexing the digits quickly or against resistance?

The flexor digitorum profundus muscle

Which flexor-pronator muscles form the:

Superficial layer of anterior forearm muscles?

From lateral to medial:
1. The pronator teres muscle
2. The flexor carpi radialis muscle
3. The palmaris longus muscle
4. The flexor carpi ulnaris muscle

Intermediate layer of anterior forearm muscles?

The flexor digitorum superficialis muscle

Deep layer of anterior forearm muscles?

1. The flexor digitorum profundus muscle
2. The flexor pollicis longus muscle
3. The pronator quadratus muscle

What is the common origin of the flexor-pronator muscle group?

The common flexor tendon, from the medial humeral epicondyle (the "funny bone")

Which three flexor-pronator muscles do not share this common origin?

The deep flexor-pronator muscles (i.e., the flexor digitorum profundus, the flexor pollicis longus, and the pronator quadratus muscles) do not arise from the common flexor tendon.

Identify the superficial flexor-pronator muscles on the following figure:

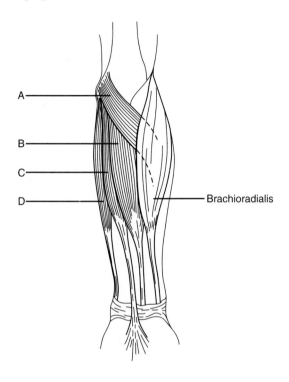

A = Pronator teres muscle
B = Flexor carpi radialis muscle
C = Palmaris longus muscle
D = Flexor carpi ulnaris muscle

**Identify the deep flexor-
pronator muscles on the
following figure:**

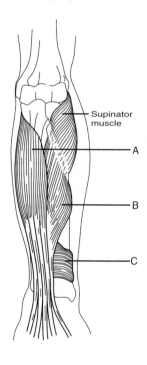

A = Flexor digitorum profundus muscle
B = Flexor pollicis longus muscle
C = Pronator quadratus muscle

Pronator teres muscle

Origin?   The medial epicondyle of the humerus

Insertion?   The lateral surface of the radius, about
                               halfway down

Innervation?   The median nerve

Action?   Pronates the hand and forearm

Flexor carpi radialis muscle

Origin?   The medial epicondyle of the humerus

Insertion?   The bases of the second and third
                               metacarpal bones

Innervation?   The median nerve

Action?	Flexes the forearm and flexes and abducts the hand
The tendon of the flexor carpi radialis muscle runs through a vertical groove in which carpal bone?	The trapezium (i.e., the first bone in the second row of carpal bones)

Palmaris longus muscle

Origin?	The medial epicondyle of the humerus
Insertion?	The flexor retinaculum (i.e., a thickened portion of the antebrachial fascia) and the palmar aponeurosis (i.e., a thickened portion of the palmar fascia)
Innervation?	The median nerve
Action?	Flexes the hand and forearm

Flexor carpi ulnaris muscle

Origin?	The medial epicondyle of the humerus, the medial olecranon of the ulna, and the posterior border of the ulna
Insertion?	The pisiform bone, the hook of the hamate, and the base of the fifth metacarpal bone
Innervation?	The ulnar nerve
Action?	Flexes and adducts the hand and flexes the forearm

Flexor digitorum superficialis muscle

Origin?	The medial epicondyle of the humerus, the coronoid process of the ulna, and the oblique line of the radius
Insertion?	The middle phalanges of the digits
Innervation?	The median nerve

| **Action?** | Flexes the proximal interphalangeal joints of the medial four digits, flexes the hand, and flexes the forearm |

Flexor digitorum profundus muscle

Origin?	The anteromedial surface of the ulna and the interosseus membrane
Insertion?	The bases of the distal phalanges of the digits
Innervation?	The ulnar and median nerves
Action?	Flexes the distal interphalangeal joints and assists in flexing the metacarpophalangeal joints and radiocarpal (wrist) joint

Flexor pollicis longus muscle

Origin?	The anterior surface of the radius, the interosseous membrane, and the coronoid process of the ulna
Insertion?	The base of the distal phalanx of the thumb
Innervation?	The median nerve
Action?	Flexes the interphalangeal joint of the thumb

Pronator quadratus muscle

Origin?	The anterior surface of the ulna
Insertion?	The anterior surface of the radius
Innervation?	The median nerve
Action?	Pronation of the forearm at the proximal and distal radioulnar joints

Posterior forearm muscles

Classify the nine extensor-supinator muscles by action. Which extensor-supinator muscles:

Extend the radiocarpal joint?

1. The extensor carpi radialis longus muscle
2. The extensor carpi radialis brevis muscle
3. The extensor carpi ulnaris muscle

Extend the metacarpo-phalangeal joints of the digits?	4. The extensor digitorum muscle 5. The extensor indicis muscle 6. The extensor digiti minimi muscle
Extend the thumb?	7. The abductor pollicis longus muscle 8. The extensor pollicis brevis muscle 9. The extensor pollicis longus muscle
Which structure prevents "bowstringing" of the extensor tendons of the wrist when the hand is hyper-extended at the carpal joint?	The extensor retinaculum (i.e., a thickening of the deep fascia of the forearm at the wrist)
What are the extensor expansions?	Located on the distal ends of the meta-carpal bones and on the phalanges, the extensor expansions are formed when the extensor tendons flatten out over the bone
What structure anchors the extensor expansions to the metacarpal bones?	The palmar ligament

Identify the labeled muscles on the following diagram of the superficial muscles of the posterior forearm:

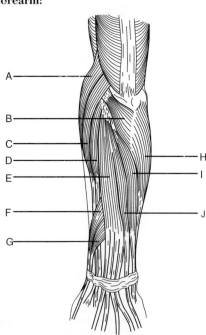

A = Brachioradialis muscle
B = Anconeus muscle
C = Extensor carpi radialis longus muscle
D = Extensor carpi radialis brevis muscle
E = Extensor digitorum muscle
F = Abductor pollicis longus muscle
G = Extensor pollicis brevis muscle
H = Flexor carpi ulnaris muscle
I = Extensor carpi ulnaris muscle
J = Extensor digiti minimi

Identify the labeled muscles on the following diagram of the deep muscles of the posterior forearm:

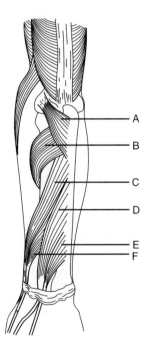

A = Anconeus muscle
B = Supinator muscle
C = Abductor pollicis longus muscle
D = Extensor pollicis longus muscle
E = Extensor indicis muscle
F = Extensor pollicis brevis muscle

What is the origin of the deep muscles of the posterior forearm?

The interosseous membrane and the posterior surfaces of the ulna and radius

Which nerve innervates all the muscles of the posterior forearm?

The radial nerve, which has superficial and deep branches

Anconeus muscle

Origin?   The lateral epicondyle of the humerus

Insertion?   The lateral surface of the olecranon and
                                     the superior part of the posterior surface
                                     of the ulna

Innervation?   The radial nerve

Action?   Assists the triceps brachii in extending the
                                     forearm at the humeroulnar joint

Brachioradialis muscle

Origin?   The proximal two thirds of the
                                     supracondylar ridge of the humerus

Insertion?   The lateral surface of the distal end of the
                                     radius

Innervation?   The radial nerve

Action?   Flexes the forearm at the humeroulnar
                                     (elbow) joint

Extensor carpi radialis brevis muscle

Origin?   The lateral epicondyle of the humerus

Insertion?   The base of the third metacarpal bone

Innervation?   The radial nerve

Action?   Extends the fingers and abducts the hand
                                     at the radiocarpal joint

Extensor carpi radialis longus muscle

Origin?   The lateral supracondylar ridge of the
                                     humerus

Insertion?   The base of the second metacarpal bone

Innervation?   The radial nerve

Action?

Extends and abducts the hand at the radiocarpal joint

Extensor carpi ulnaris muscle

Origin?

The lateral epicondyle and posterior surface of the ulna

Insertion?

The base of the fifth metacarpal bone

Innervation?

The radial nerve

Action?

Extends and adducts the hand at the radiocarpal joint

Extensor digiti minimi muscle

Origin?

The common extensor tendon and the interosseus membrane

Insertion?

The extensor expansions and the bases of the middle and distal phalanges

Innervation?

The radial nerve

Action?

Extends the fifth digit (i.e., the "little" finger)

Extensor digitorum muscle

Origin?

The lateral epicondyle of the humerus

Insertion?

The extensor expansions and the bases of the middle and distal phalanges

Innervation?

The radial nerve

Action?

Extends the fingers at the metacarpophalangeal joints and the hand at the radiocarpal joint

Abductor pollicis longus muscle

Origin?

The interosseus membrane and the posterior surfaces of the radius and ulna

Insertion?

The base of the first metacarpal bone

Innervation?   The radial nerve

Action?   Abducts the thumb at the
                                    carpometacarpal joint and the hand at the
                                    radiocarpal joint

Extensor pollicis brevis muscle

Origin?   The interosseus membrane and posterior
                                    surface of the radius

Insertion?   The base of the proximal phalanx of the
                                    thumb

Innervation?   The radial nerve

Action?   Extends the proximal phalanx of the
                                    thumb and abducts the hand at the
                                    radiocarpal joint

Extensor pollicis longus muscle

Origin?   The interosseus membrane and the
                                    posterior surface of the ulna

Insertion?   The base of the distal phalanx of the
                                    thumb

Innervation?   The radial nerve

Action?   Extends the distal phalanx of the thumb
                                    and abducts the hand at the radiocarpal
                                    joint

Extensor indicis muscle

Origin?   The interosseus membrane and the
                                    posterior surface of the ulna

Insertion?   The extensor expansion of the second
                                    digit (the index finger)

Innervation?   The radial nerve

Action?   Extends the second digit

Supinator muscle

Origin?   The lateral epicondyle of the humerus,
                                    the radial collateral and annular

ligaments, the supinator fossa, and the crest of the ulna

Insertion?

The lateral, posterior, and anterior surfaces of the proximal third of the radius

Innervation?

The radial nerve

Action?

Supination of the forearm and hand at the radioulnar joint

VASCULATURE OF THE FOREARM

Which artery in the forearm is often used to palpate an arterial pulse?

The radial artery, which lies lateral to the tendon of the flexor carpi radialis muscle

The ulnar artery accompanies which nerve on its course between the two heads of the flexor digitorum superficialis muscle?

The median nerve (note the ulnar artery subsequently runs medially to join the ulnar nerve)

The ulnar artery lies between which two muscles?

The flexor digitorum superficialis and the flexor digitorum profundus

The common interosseous artery (a branch of the ulnar artery) divides into which two arteries?

The anterior interosseous artery and the posterior interosseous artery

Before anastomosing with the posterior interosseous artery, which fascial membrane does the anterior interosseous artery pierce?

The interosseous membrane

INNERVATION OF THE FOREARM

Describe the course of the ulnar nerve at the elbow.

The ulnar nerve passes between the medial epicondyle of the humerus (the "funny bone") and the olecranon of the ulna to enter the forearm.

**The heads of which flexor-
pronator muscle are split
by the:**

    Median nerve?   The pronator teres muscle

    Ulnar nerve?   The flexor carpi ulnaris muscle

JOINTS AND LIGAMENTS OF THE FOREARM

**Describe the proximal
radioulnar joint.**

The proximal radioulnar joint is a pivot-type synovial joint in which the head of the radius articulates with the radial notch of the ulna.

**Describe the distal radio-
ulnar joint.**

The distal radioulnar joint is a pivot-type synovial joint in which the head of the ulna articulates with the ulnar notch of the radius.

**The proximal and distal
radioulnar joints allow for
which movements of the
forearm?**

Pronation and supination

**Which muscles are
involved in:**

    Pronation of the forearm   The pronator quadratus and pronator teres muscles

    Supination of the forearm?   The supinator and biceps brachii muscles

WRIST (CARPUS)

**Describe the radiocarpal
(wrist) joint.**

The radiocarpal joint is a condyloid joint where the radius and articular disk articulate with the scaphoid, lunate, and triquetral bones.

**Which movements are
possible at the radiocarpal
joint?**

Flexion and extension, abduction and adduction, and circumduction

**Name the four ligaments
associated with the radiocarpal joint.**

1. Anterior ligament
2. Posterior ligament
3. Radial collateral ligament
4. Ulnar collateral ligament

**Describe the intercarpal
joint.**

The intercarpal joint is the synovial joint between the proximal and distal rows of carpal bones.

Which movements are allowed by the intercarpal joint?	Flexion and abduction of the hand
Which structure converts the carpal groove into a "tunnel?"	The flexor retinaculum
How is the flexor retinaculum formed?	The deep fascia of the forearm thickens anteriorly at the wrist.
Describe the attachments of the flexor retinaculum.	The scaphoid and trapezium (i.e., the lateral-most carpal bones) laterally and the pisiform and hamate (the medial-most carpal bones) medially
Name the structures that enter the palm superficial to the flexor retinaculum.	**UP UP** above the flexor retinaculum **U**lnar nerve **P**almaris longus tendon **U**lnar artery **P**almar cutaneous branch of the median nerve
What are the anterior and posterior boundaries of the carpal tunnel?	The flexor retinaculum anteriorly and the carpal bones posteriorly
What structures pass through the carpal tunnel?	1. Median nerve 2. Flexor pollicis longus tendon 3. Flexor digitorum profundus tendons (four) 4. Flexor digitorum superficialis tendons (four)
What structure, deep to the flexor retinaculum, encloses the flexor digitorum profundus and flexor digitalis superficialis tendons?	The common flexor synovial sheath
How is the extensor retinaculum formed?	The deep fascia of the forearm thickens posteriorly at the wrist.
Describe the attachments of the extensor retinaculum.	The styloid process of the ulna medially, and the triquetrum and pisiform bones laterally
Which nerve crosses the extensor retinaculum?	The superficial branch of the radial nerve

HAND (MANUS)

ANATOMIC SNUFF BOX

What are the boundaries of the anatomic snuff box:

    Posteriorly?　　　　　　The tendon of the extensor pollicis longus muscle

    Anteriorly?　　　　　　The tendons of the extensor pollicis brevis and abductor pollicis longus muscles

    Proximally?　　　　　　The styloid process of the radius

What forms the floor of the anatomic snuff box?　　　　The scaphoid and trapezium bones

Which artery lies in the anatomic snuff box?　　　　The radial artery

Tenderness over the anatomic snuff box suggests fracture of which bone?　　　　The scaphoid bone

JOINTS AND LIGAMENTS OF THE HAND

Which ligaments support the metacarpophalangeal joints?　　　　Each joint is supported by one palmar ligament and two collateral ligaments.

What is the clinical name for enlargement of the:

    Proximal interphalangeal joints?　　　　Bouchard's nodes (characteristic in rheumatoid arthritis)

    Distal interphalangeal joints?　　　　Heberden's nodes (characteristic in osteoarthritis)

FASCIAE AND MUSCLES OF THE HAND

Name the four fascial compartments in the hand.
1. Thenar compartment
2. Adductor compartment
3. Hypothenar compartment
4. Central compartment

What is the fibrous structure that overlays the tendons in the palm?　　　　The palmar aponeurosis

Palmaris brevis muscle

Origin?

The flexor retinaculum and palmar aponeurosis

Insertion?

The skin of the medial palm

Innervation?

The ulnar nerve

Action?

Wrinkles the skin of the medial palm

Intrinsic hand muscles

Name the four groups of intrinsic muscles of the hand.

1. Thenar muscles
2. Adductor pollicis muscle
3. Hypothenar muscles
4. Short muscles (i.e., the lumbricals and interossei)

Thenar muscles

Name the three thenar muscles.

1. Abductor pollicis brevis muscle
2. Flexor pollicis brevis muscle
3. Opponens pollicis muscle

What is the major action of the thenar muscles?

Abduction, flexion, and opposition of the carpometacarpal joint of the thumb

Which nerve innervates the thenar muscles?

The median nerve

Abductor pollicis brevis muscle

Origin?

The flexor retinaculum, the scaphoid bone, and the trapezium bone

Insertion?

The base of the proximal phalanx of the thumb

Action?

Abducts the thumb

Flexor pollicis brevis muscle

Origin?

The flexor retinaculum and the trapezius bone

Insertion?

The base of the proximal phalanx of the thumb

Action?

Flexes the thumb

Opponens pollicis muscle

Origin?   The flexor retinaculum and the trapezius
                                                bone

Insertion?   The first metacarpal

Action?   Opposes the thumb to the other digits

Adductor pollicis muscle

Origin?   The capitate bone, the bases of the
                                                second and third metacarpals, and the
                                                palmar surface of the third metacarpal

Insertion?   The proximal phalanx of the thumb

Innervation?   The ulnar nerve

Action?   Adducts the thumb at the
                                                carpometacarpal joint

**What structure separates   The radial artery (as it forms the deep
the two heads of the   palmar arch)
adductor pollicis muscle?**

Hypothenar muscles

**Name the three hypothenar   1. Abductor digiti minimi muscle
muscles.**   2. Flexor digiti minimi brevis muscle
                                                3. Opponens digiti minimi muscle

**Which nerve innervates the   The ulnar nerve
hypothenar muscles?**

Abductor digiti minimi muscle

Origin?   The pisiform bone and the tendon of the
                                                flexor carpi ulnaris muscle

Insertion?   The proximal phalanx of the fifth digit

Action?   Abducts the fifth digit

Flexor digiti minimi brevis muscle

Origin?   The flexor retinaculum and the hook of
                                                the hamate

Insertion?   The proximal phalanx of the fifth digit

Action?	Flexes the proximal phalanx of the fifth digit

Opponens digiti minimi muscle

Origin?	The flexor retinaculum and the hook of the hamate
Insertion?	The fifth metacarpal
Action?	Opposes the fifth digit

Short muscles

What are the short muscles of the hand?	1. The lumbricals 2. The interossei (dorsal and palmar)

Lumbricals

Origin?	The lateral sides of the tendons of the flexor digitorum profundus muscle
Insertion?	The lateral sides of the extensor expansions
Innervation?	The median nerve innervates the two lateral lumbricals, and the ulnar nerve innervates the two medial lumbricals.
Action?	Flexion of the metacarpophalangeal joints and extension of the proximal and distal interphalangeal joints

Interossei

Origin?	**Dorsal interossei:** The sides of the metacarpals **Palmar interossei:** The medial side of the second metacarpal and the lateral sides of the fourth and fifth metacarpals
Insertion?	**Dorsal interossei:** The lateral sides of the bases of the proximal phalanges and the extensor expansions **Palmar interossei:** The bases of the proximal phalanges and the extensor expansions
Innervation?	The ulnar nerve

What are the actions of the:

Dorsal interossei muscles?
1. Abduction of the digits from the axial line (i.e., the third digit); remember **DAB** (**D**orsal **AB**ducts)
2. Flexion of the metacarpophalangeal joints
3. Extension of the interphalangeal joints

Palmar interossei muscles?
1. Adduction of the digits to the axial line; remember **PAD** (**P**almar **AD**ducts)
2. Flexion of the metacarpophalangeal joints
3. Extension of the interphalangeal joints

VASCULATURE AND INNERVATION OF THE HAND

In the hand, the radial artery divides into which vessels?

The princeps pollicis artery and the deep palmar arch

The princeps pollicis artery divides into which vessels?

Two proper digital arteries

Identify each region on the following figure by the nerve that innervates it:

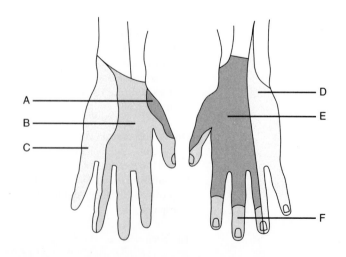

A = Radial nerve
B = Median nerve
C = Ulnar nerve
D = Ulnar nerve
E = Radial nerve
F = Median nerve

POWER REVIEW

PECTORAL GIRDLE AND SHOULDER

Which six muscles pass from the scapula to the humerus and act on the shoulder joint?

1. Supraspinatus muscle
2. Infraspinatus muscle
3. Teres minor muscle
4. Subscapularis muscle
5. Deltoid muscle
6. Teres major muscles

Which four muscles aid in the medial rotation of the humerus?

1. Deltoid muscle
2. Subscapularis muscle
3. Teres major muscle
4. Latissimus dorsi muscle

Which two muscles rotate the humerus laterally?

1. Infraspinatus muscle
2. Teres minor muscle

What are the four "rotator cuff" muscles?

SITS
Supraspinatus muscle
Infraspinatus muscle
Teres minor muscle
Subscapularis muscle

Which muscles insert on the:

Greater tubercle of the humerus?

The supraspinatus, infraspinatus, and teres minor muscles (i.e., the posterior rotator cuff muscles)

Lesser tubercle of the humerus?

The subscapularis muscle (i.e., the anterior rotator cuff muscle)

Of the four rotator cuff muscles, which does not rotate the humerus?

The supraspinatus muscle (the infraspinatus and teres minor muscles rotate the humerus laterally, while the subscapularis muscle is involved in medial rotation)

Identify the labeled muscular attachments on the following figure, showing the posterior surfaces of the clavicle, scapula, and proximal humerus:

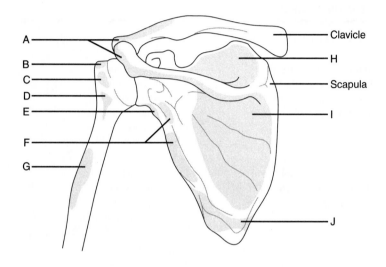

A = Origin of the deltoid muscle
B = Insertion of the supraspinatus muscle
C = Insertion of the infraspinatus muscle
D = Insertion of the teres minor muscle
E = Origin of the triceps muscle, long head
F = Origin of the teres minor muscle
G = Insertion of the deltoid muscle
H = Origin of the supraspinatus muscle
I = Origin of the infraspinatus muscle
J = Origin of the teres major muscle

AXILLA

The axillary artery extends from where to where?	The lateral border of the first rib to the inferior border of the teres major muscle
Which muscle divides the axillary artery into three parts?	The pectoralis minor muscle

ARM

Which nerve and vessel run in the spiral groove on the back of the humerus, and are therefore susceptible to damage following fracture of the humeral shaft?	The radial nerve and the deep brachial artery
Describe the four muscles of the arm in terms of their fascial compartments.	**Anterior fascial compartment:** Biceps brachii, brachialis, and coraco-brachialis muscles (flexors) **Posterior fascial compartment:** Triceps brachii muscle (extensor)

Which nerve innervates the:

Flexors of the upper limb?	The musculocutaneous nerve
Extensors of the upper limb?	The radial nerve
Which muscle is the main flexor of the humeroulnar joint?	The brachialis muscle, **not** the biceps brachii muscle!
Which muscle is the main extensor of the humero-ulnar joint?	The triceps brachii muscle
Weakened flexion of the humeroulnar joint and supination of the forearm suggests damage to which nerve?	The musculocutaneous nerve

ELBOW REGION AND FOREARM

Which nerve passes behind the medial humeral epi-condyle ("funny bone")?	The ulnar nerve

All of the flexor muscles in the forearm are innervated by the median or ulnar nerves except for which one?

The brachioradialis muscle—despite being a flexor, the brachioradialis muscle is supplied by the radial nerve; this means that it is the one exception to the rule that the radial nerve is the "nerve of the extensors."

Which nerve innervates all of the posterior muscles of the forearm?

The radial nerve

WRIST AND HAND

What carpal bone is the most commonly fractured?

The scaphoid bone; fracture causes tenderness in the anatomic snuff box

Which structures pass through the carpal tunnel?

The median nerve (this is the nerve that is compressed in carpel tunnel syndrome), the flexor pollicis longus tendon, the four flexor digitorum profundus tendons, and the four flexor digitorum superficialis tendons

Which artery lies in the anatomic snuff box?

The radial artery

In the hand, which nerve innervates the:

Thenar muscles?

Medial nerve

Hypothenar muscles?

Ulnar nerve

8

The Thorax

THORACIC WALL

Which bones form the skeleton of the thoracic wall?

1. Vertebrae T1–T12
2. Ribs 1–12 and their associated costal cartilages
3. The sternum

The sternum is composed of which three segments?

The manubrium, the body, and the xiphoid process

Which bones form the borders of the superior thoracic aperture?

Vertebra T1, rib 1, and the manubrium of the sternum

The sternal angle (manubriosternal joint) marks the location of which three things?

1. The beginning and ending of the aortic arch
2. The carina of the lungs (i.e., the bifurcation of the trachea into the left and right main stem bronchi)
3. The separation of the superior and inferior mediastinum [draw a horizontal line from the sternal angle (anteriorly) to the vertebrae T4–T5 intervertebral disk (posteriorly)]

What is contained in the costal grooves, immediately inferior to each rib?

Neurovascular bundles (**VAN: V**ein, **A**rtery, **N**erve)

How many intercostal spaces are there, and what do they contain?

Each contains a neurovascular bundle and three muscle layers.

Name the three intercostal muscle layers.

1. External intercostal
2. Internal intercostal
3. Innermost intercostal

The anterior intercostal arteries are branches of which artery?	The internal thoracic (internal mammary) artery, the first branch of the subclavian artery
The posterior intercostal arteries are branches of which artery?	The descending thoracic aorta
Which muscle connects the sternum to the ribs internal to the internal thoracic artery?	The transversus thoracis muscle
What are the terminal branches of the internal thoracic artery?	The superior epigastric artery and the musculophrenic artery
Which nerves innervate the body wall?	The intercostal nerves (i.e., the ventral primary rami of spinal cord levels T1–T11) are both motor and sensory to the body wall and sensory to the parietal pleura and the periphery of the diaphragm.

Which dermatome supplies sensation to the:

Nipple?	T4
Umbilicus?	T10

THORACIC CAVITY

The thoracic cavity is divided into how many compartments?	Three—two lateral compartments (which contain the lungs and pleurae) and a central compartment, known as the mediastinum, which contains the other thoracic structures

MEDIASTINUM

What are the boundaries of the mediastinum:

Anteriorly?	The sternum
Posteriorly?	The thoracic vertebrae
Laterally?	The pleura

Superiorly?	The superior thoracic aperture
Inferiorly?	The diaphragm
The superior and inferior mediastinum are separated by a plane passing between which structures?	The sternal angle and the T4–T5 intervertebral disk

Superior mediastinum

What structures are contained within the superior mediastinum?

1. The thymus gland
2. The trachea
3. The upper third of the esophagus
4. The thoracic duct
5. The vagi and phrenic nerves and the recurrent laryngeal nerve
6. The great vessels

What are the first branches off of the aorta?	The right and left coronary arteries
What is the first branch off the *arch* of the aorta?	The brachiocephalic artery
The brachiocephalic artery gives off which two major branches?	The right common carotid artery and the right subclavian artery
What is the second branch off the arch of the aorta?	The left common carotid artery
What is the third branch off the arch of the aorta?	The left subclavian artery
Which structure hooks around the aortic arch just lateral to the ligamentum arteriosum?	The left recurrent laryngeal nerve
Which two veins join to form the superior vena cava?	The left and right brachiocephalic veins
Which two veins join to form the brachiocephalic vein?	The internal jugular vein and the subclavian vein
Where does the thoracic duct empty?	Into the left internal jugular or left subclavian vein, just before these veins join each other

Identify the following structures on this transverse cut just above the aortic arch, as if looking at a computed tomography (CT) or magnetic resonance imaging (MRI) scan (i.e., from the feet up; the left side of the body is the right side of the page):

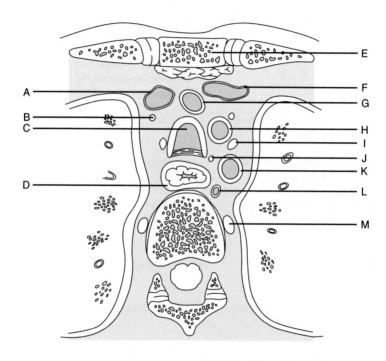

A = Right brachiocephalic vein
B = Phrenic nerve
C = Trachea
D = Esophagus
E = Manubrium
F = Left brachiocephalic vein
G = Brachiocephalic artery
H = Left common carotid artery
I = Vagus nerve
J = Left recurrent laryngeal nerve
K = Left subclavian artery
L = Thoracic duct
M = Sympathetic trunk

Inferior mediastinum

The inferior mediastinum is divided into which three compartments?

Anterior, middle, and posterior compartments

What is located in the anterior mediastinum?

A portion of the thymus gland

List seven structures contained in the posterior mediastinum.

1. The lower two thirds of the esophagus
2. The anterior and posterior esophageal plexuses
3. The descending aorta
4. The thoracic duct
5. Splanchnic nerves
6. The azygos vein
7. The hemiazygos vein

Which artery supplies the:

    Upper third of the esophagus?

The inferior thyroid artery

    Middle third of the esophagus?

The bronchial and esophageal arteries, branches of the descending aorta

    Lower third of the esophagus?

The left gastric and left inferior phrenic arteries

What structure is contained within the middle mediastinum?

The heart

HEART

What are the three layers of cardiac tissue?

1. Epicardium
2. Myocardium
3. Endocardium

What is the apex of the heart?

The pointed portion of the heart formed by the projection of the left ventricle downward, forward, and to the left

What is the base of the heart?

The posterosuperior portion of the heart formed primarily by the left atrium

What is the margin between the atria and ventricles called?

The coronary sulcus

What is the margin between the two ventricles called?	The interventricular groove
What are the two types of pericardium that surround the heart?	1. Fibrous pericardium (the tough outer coat) 2. Serous pericardium (consisting of the visceral serous pericardium, on the surface of the heart itself, and the parietal serous pericardium, on the interior of the fibrous pericardium)
Where is the pericardial fluid located?	In the pericardial cavity, a potential space between the visceral serous pericardium and the parietal serous pericardium

CHAMBERS

What are the muscular ridges on the surfaces of the ventricular walls called?	Trabeculae carneae
What are the two portions of the ventricular septum?	The muscular portion and the membranous portion
What is the trabeculated muscle on the interior surface of the right atrium called?	Pectinate muscle
What is the smooth posterior wall of the right atrium called?	The sinus venarum
What is the blood supply of the sinus venarum?	The superior vena cava, inferior vena cava, and coronary sinus
What is the small depression in the interatrial septum (on the right atrial side) called?	The fossa ovalis
The fossa ovalis is a remnant of which embryologic structure?	The foramen ovale

VALVES

Which two valves are the:

Inflow valves of the ventricles?	1. The tricuspid (right atrioventricular) valve 2. The mitral (left atrioventricular) valve

Outflow valves of the ventricles?	1. The pulmonic valve 2. The aortic valve
Describe the flow of blood through the heart.	Blood flows from the right atrium to the right ventricle through the tricuspid valve. From the right ventricle, blood flows to the pulmonary artery through the pulmonic valve and through the lungs into the left atrium. From the left atrium, blood passes to the left ventricle through the mitral valve, and then to the aorta through the aortic valve.
What are the valves formed from?	Thin sheets of fibrous tissue called cusps (leaflets)
How many leaflets does the mitral valve have?	Two
What is the orientation of the two leaflets of the mitral valve?	Anterior and posterior
How many leaflets does the tricuspid valve have?	Three
What is the orientation of the three leaflets of the tricuspid valve?	Anterior, posterior, and septal (medial)
Each mitral and tricuspid valve leaflet is attached to how many papillary muscles?	Two
What structures connect the mitral and tricuspid valve leaflets to the papillary muscles?	Chordae tendineae
What is the function of the papillary muscles?	They help hold the ventricular inflow valves closed during ventricular ejection (contraction)

Does the pulmonic valve lie anterior or posterior to the aortic valve?	Anterior to it
How many leaflets does the pulmonic valve have?	Three
What is the orientation of the three leaflets of the pulmonic valve?	Right, left, and anterior
How many leaflets does the aortic valve have?	Three
What is the orientation of the three leaflets of the aortic valve?	Right, left, and posterior

CONDUCTING SYSTEM

Where is the sinoatrial node located?	At the junction of the superior vena cava and the right atrium
What constitutes the blood supply of the sinoatrial node?	The sinoatrial nodal artery, which in 60% of patients, is a branch of the right coronary artery
Where is the atrioventricular node located?	In the atrial septum, near where the coronary sinus enters the right atrium
What constitutes the blood supply of the atrioventricular node?	The atrioventricular node is almost always supplied by the atrioventricular nodal artery, a branch of the right coronary artery.
What are the modified myocardial cells that are responsible for the rapid conduction of electrical impulses through the heart called?	Purkinje fibers

Identify the structures that form the heart's conducting system on the following figure:

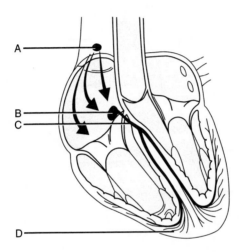

A = Sinoatrial node
B = Atrioventricular node
C = Atrioventricular bundle (of His)
D = Purkinje fibers

What is the path of electrical stimulation in the normally functioning heart?

The impulse arises spontaneously in the **sinoatrial node** and travels across the atria, causing them to contract. The impulse stimulates the **atrioventricular node,** which distributes the impulse through the **atrioventricular bundle (of His).** The atrioventricular bundle (of His) has right and left branches consisting of collections of specialized conducting myocytes called **Purkinje fibers.** From the Purkinje fibers, the impulse passes to the **ventricular myocytes,** resulting in ventricular contraction.

VASCULATURE

Identify the lettered structures corresponding to the arterial supply of the heart on this sternocostal view:

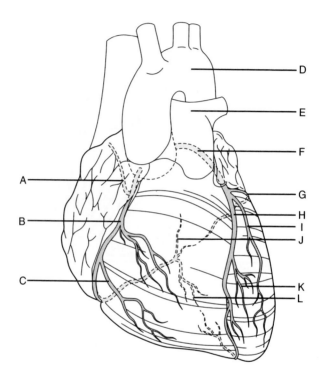

A = Sinoatrial nodal artery
B = Right coronary artery
C = Right marginal artery
D = Aortic arch
E = Pulmonary trunk
F = Left main coronary artery
G = Circumflex artery
H = Anterior interventricular (left anterior descending) artery
I = Left marginal artery
J = Atrioventricular nodal artery
K = Diagonal artery
L = Posterior interventricular (posterior descending) artery

Where do the right and left main coronary arteries arise?

From the sinuses of Valsalva, which are located within the right and left leaflets of the aortic valve

The right coronary artery generally gives rise to which two major branches?

1. The marginal artery
2. The posterior interventricular (posterior descending) artery

The anterior right atrial branch of the right coronary artery gives rise to which small, yet very important, branch?

The sinoatrial nodal artery

The left main coronary artery gives rise to which two major branches?

1. The circumflex artery
2. The anterior interventricular (left anterior descending) artery

What are the two major types of branches of the left anterior descending artery?

The diagonal branches and the perforating branches

Which structure is supplied by the diagonal branches and the perforating branches?

The anterior two thirds of the interventricular septum

Identify the shaded structures corresponding to the venous drainage of the heart, as well as the other labeled structures, on the following sternocostal view of the heart:

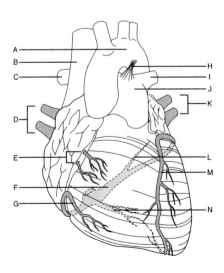

A = Aortic arch
B = Superior vena cava
C = Right pulmonary artery
D = Right pulmonary veins
E = Anterior cardiac veins
F = Coronary sinus
G = Small cardiac vein
H = Ligamentum arteriosum
I = Left pulmonary artery
J = Pulmonary trunk
K = Left pulmonary veins
L = Oblique cardiac vein
M = Great cardiac vein
N = Middle cardiac vein

All of the major cardiac veins terminate in which structure?

The coronary sinus, located in the posterior coronary sulcus

Which cardiac vein travels with the:

Posterior descending artery?

The middle cardiac vein

Right marginal artery?

The small cardiac vein

Which vein travels in the anterior interventricular sulcus with the left anterior descending artery?

The great cardiac vein

Where is the coronary vein located?

The coronary vein is another name for the left gastric vein, which takes a circular course along the lesser curvature of the stomach to anastomose with the right gastric vein, forming a ring (*corona* = crown or ring). Both the left and right gastric veins drain into the portal vein.

INNERVATION

What is the efferent innervation of the heart?

The vagus nerve provides parasympathetic innervation to the sinoatrial and atrioventricular nodes and the coronary arteries.

What is the afferent innervation of the heart?

The vagus nerve carries afferent impulses to the cardiovascular reflex centers. Pain (such as that from ischemia or infarction) travels with the sympathetic fibers.

LUNGS

Name the two fissures of the right lung.	Oblique (major) and horizontal (minor)
What is the fissure of the left lung?	The oblique (major) fissure
Name the three lobes of the right lung.	1. Right upper lobe 2. Right middle lobe 3. Right lower lobe
Name the two lobes of the left lung.	1. Left upper lobe 2. Left lower lobe
What is the lingula?	The lingula (from Latin *lingua,* meaning tongue) is a small extension of the left upper lobe; you can think of it as being the left-sided equivalent of a middle lobe.
Name six components of the root of the lung.	1. Primary (main stem) bronchus 2. Pulmonary artery 3. Superior and inferior pulmonary veins 4. Bronchial artery and vein 5. The pulmonary nerve plexus 6. Lymphatics
Which of these structures is posterior?	The primary (main stem) bronchus
Which of these structures is anterior and inferior?	The pulmonary veins
How does the location of the pulmonary artery differ from the right to the left?	On the left, the pulmonary artery is superior to the primary (main stem) bronchus and the pulmonary vein, whereas on the right, the pulmonary artery is between the primary (main stem) bronchus and the pulmonary vein.
Describe the ten terms used to describe the airways as they get progressively smaller.	1. Trachea 2. Primary (main stem) bronchus 3. Secondary (lobar) bronchus 4. Tertiary (segmental) bronchus 5. Bronchiole 6. Terminal bronchiole 7. Respiratory bronchiole

8. Alveolar duct
9. Alveolar sac
10. Alveolus
Note that the terminal bronchiole is not, in fact, the end of the line!

PLEURAE

What are the two major pleurae of the lungs?

The visceral pleura (which invests the lungs) and the parietal pleura (which lines the thoracic wall)

What is the pleural cavity?

A potential space between the parietal and visceral pleura that contains a small amount of pleural fluid

Name the four types of parietal pleura (as designated by location).

1. Mediastinal pleura
2. Costal pleura
3. Diaphragmatic pleura
4. Cervical pleura (pleural cupula)

The sleeve of pleura just inferior to the root of the lung is called what?

The pulmonary ligament

Which nerves innervate the parietal pleura?

The intercostal nerves innervate the costal and peripheral diaphragmatic pleurae, and the phrenic nerves innervate the mediastinal and central diaphragmatic pleurae.

The visceral pleura receives somatic afferents from which nerve?

None! There is no somatic sensation of the visceral pleura. Instead, autonomic innervation sensitizes the visceral pleura to stretch.

VASCULATURE

The bronchi and pulmonary connective tissue receive oxygenated blood from which arteries?

The bronchial arteries, branches of the thoracic aorta

What is the venous drainage of the lung?

The four pulmonary veins (carrying oxygenated blood) drain into the left atrium. In addition, there are left and right bronchial veins, which drain deoxygenated blood from the tissue near the hilum of the lung into the azygos and hemiazygos veins.

Describe the three main components of the azygos venous system.

1. The **azygos vein** proper forms a bridge between the superior vena cava and the inferior vena cava and accepts flow from the intercostal veins. Like the vena cavae, the azygos vein runs on the right side.
2. The **hemiazygos vein** drains the inferior left side and empties into the azygos vein.
3. The **accessory hemiazygos vein** drains the superior left side before emptying into the azygos vein.

What is the name of the lymph nodes found:

Within the pulmonary parenchyma?

Pulmonary nodes

At the hilum?

Bronchopulmonary nodes

At the carina?

Tracheobronchial nodes

Tracheobronchial nodes on the right side drain to which structure?

The right thoracic duct

On the left side?

The left thoracic duct

DIAPHRAGM

What is the inferior boundary of the thoracic cavity?

The diaphragm

What three arteries supply blood to the diaphragm?

1. Inferior phrenic artery
2. Musculophrenic artery
3. Pericardiacophrenic artery

Which nerve innervates the diaphragm?

The phrenic nerve

Where does the phrenic nerve travel through the thorax?

Just on the outside of the pericardium

Which vessels does the phrenic nerve travel with through the thorax?	The pericardiacophrenic artery and vein, small branches of the internal thoracic artery
What are the three openings in the diaphragm that allow structures to pass from the thoracic cavity into the abdomen?	1. Aortic hiatus 2. Esophageal hiatus 3. Caval foramen
Which three structures pass through the aortic hiatus?	1. Aorta 2. Azygos vein 3. Thoracic duct
The aortic hiatus is located at which vertebral level?	T12
Which structures form the aortic hiatus?	The right and left crura and the median arcuate ligament
Aside from the esophagus, name three structures that pass through the esophageal hiatus.	1. Vagal trunks 2. Esophageal branches of the left gastric vessels 3. Lymphatics from the inferior third of the esophagus
The esophageal hiatus is located at which vertebral level?	T10
Which structures pass through the caval foramen?	1. Inferior vena cava 2. Terminal branches of the right phrenic nerve
The caval foramen is located at which vertebral level?	T8

BREAST

In men, the nipple typically lies in which intercostal space?	The fourth
In women, which structures does milk drain through to exit the breast?	Each lobe drains into a lactiferous duct, which in turn drains into a lactiferous sinus. Milk drains from the lactiferous sinus through the nipple to the outside of the body.

What are the ligaments within the breast called?	Cooper's ligaments, or the suspensory ligaments of the breast
What are the three major arteries that supply blood to the breast?	1. Perforating branches of the internal mammary artery 2. Lateral mammary branches of the lateral thoracic artery 3. Pectoral branches of the thoracoacromial artery
Which arteries make minor contributions to the blood supply of the breast?	The subscapular artery and the lateral intercostal perforating arteries
What percentage of breast lymph node drainage is lateral?	Approximately 75%
Where do the lateral lymph nodes of the breast drain to?	Most drain to the axillary lymph nodes, first to the pectoral group (deep to the pectoralis major muscle along the inferior border of the pectoralis minor muscle), and then to the superficial apical group.
Where do the medial lymph nodes of the breast drain to?	The parasternal lymph nodes, which run with the internal thoracic vein

POWER REVIEW

How do the neurovascular bundles run along the thoracic cage?	In the costal grooves beneath the ribs
What are the first branches of the aorta?	The right and left coronary arteries
What are the branches of the aortic arch?	The brachiocephalic artery, the left common carotid artery, and the left subclavian artery
Which is the most anterior of the great vessels in the midline?	The left brachiocephalic vein, which crosses the midline to join the right brachiocephalic vein and form the superior vena cava
What are the two major arteries of the heart, and which parts of the heart do they supply?	1. The **left main coronary artery** gives off two branches. The anterior interventricular (left anterior descending) artery supplies the

anterior portion of the left ventricle and the anterior two thirds of the interventricular septum, and the left circumflex coronary artery supplies the lateral wall of the left ventricle.

2. The **right coronary artery** supplies the posterior wall of the left ventricle and the posterior third of the interventricular septum through its posterior interventricular branch (i.e., the posterior descending artery).

Which artery supplies the sinoatrial node?

The sinoatrial nodal artery, usually a branch of the right coronary artery

Which heart valve has two cusps?

The mitral valve

What forms the base of the heart?

The left atrium

What are the fissures of the left and right lungs?

Left: Oblique (major) fissure (divides the lung into two lobes)

Right: Horizontal (minor) and oblique (major) fissures (divide the lung into three lobes)

What is the reflection of pleura inferior to the root of the lung called?

The pulmonary ligament

Which is posterior in the root of the lung, the primary (main stem) bronchus or the pulmonary artery?

The primary (main stem) bronchus

What are the only arteries in the body to carry deoxygenated blood?

The pulmonary arteries

Which vessels form the dual blood supply to the lungs?

The pulmonary arteries (which carry deoxygenated blood) and the bronchial arteries (direct branches of the aorta which carry oxygenated blood)

At which levels do the three major structures pass through the diaphragm?

1. Inferior vena cava: T8
2. Esophagus: T10
3. Aorta: T12

Think **"A, E, I"—alphabetically up the spine)**

The esophageal hiatus is formed by which portion of the diaphragm?

The right crus

Innervation of the diaphragm is by which spinal nerves?

C3, C4, and C5 "keep the diaphragm alive"

9   The Abdomen

**Identify the bony landmarks
of the abdomen on the following
figure:**

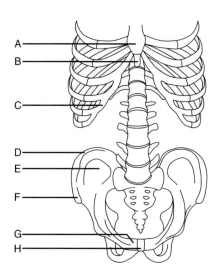

A = Body of sternum
B = Xiphoid process
C = Costal cartilages
D = Iliac crest
E = Iliac fossa
F = Anterior superior iliac spine
G = Pubic tubercle
H = Pubic symphysis

Identify the nine abdominal regions on the following figure:

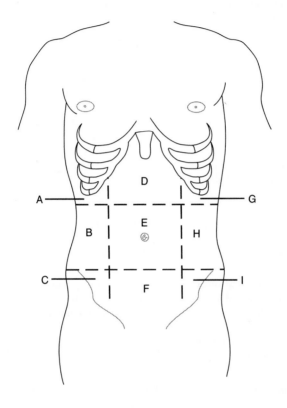

A = Right hypochondriac region
B = Right lumbar region
C = Right inguinal region
D = Epigastric region
E = Umbilical region
F = Suprapubic region
G = Left hypochondriac region
H = Left lumbar region
I = Left inguinal region
Note that in practice, clinicians more often refer to the abdomen in terms of four quadrants, delineated by horizontal and vertical planes through the umbilicus.

ABDOMINAL WALL

ANTERIOR ABDOMINAL WALL

Name the nine layers of the anterior abdominal wall, from superficial to deep.

1. Skin
2. Camper's fascia (fatty layer of the superficial fascia)
3. Scarpa's fascia (membranous layer of the superficial fascia)
4. External oblique muscle
5. Internal oblique muscle
6. Transversus abdominis muscle
7. Transversalis fascia
8. Preperitoneal fatty tissue
9. Peritoneum

What represents the continuation of Scarpa's fascia in the:

Thigh? The fascia lata

Perineum? Colles' fascia

Name the three flat muscles of the anterolateral abdominal wall.

1. External oblique muscle
2. Internal oblique muscle
3. Transversus abdominis muscle

Describe the direction that the fibers course in the:

External oblique muscle Inferomedially ("*hands in pockets* because it's cold *outside*")

Internal oblique muscle Inferolaterally (at right angles to the fibers of the external oblique muscle)

Transversus abdominis muscle Transversely (horizontally)

Where do the internal oblique and transversus abdominis muscles insert? In textbooks, the internal oblique and transversus abdominis muscles insert into the pubic tubercle via the conjoint tendon. In actuality, this structure rarely exists; more commonly, the transversus abdominis aponeurosis alone attaches to the pubic crest.

Name the strap-like vertical muscle of the anterolateral abdominal wall.

The rectus abdominis muscle

What structure encloses most of the rectus abdominis muscle?

The rectus sheath

What forms the rectus sheath?

The aponeuroses of the external oblique, internal oblique, and transversus abdominis muscles

In addition to the rectus abdominis muscle, what structures are enclosed by the rectus sheath?

1. The pyramidalis muscle (in 80% of people)
2. The superior and inferior epigastric arteries and veins
3. Lymphatic vessels
4. The T7–T12 ventral primary rami

Which vessel gives rise to the:

Superior epigastric artery?

Internal thoracic artery

Inferior epigastric artery?

External iliac artery

Identify the lettered structures on the following figure of the anterior abdominal wall:

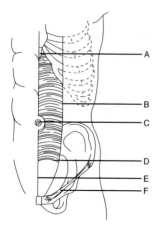

A = Xiphoid process
B = Linea semilunaris
C = Umbilicus
D = Arcuate line
E = Linea alba
F = Inguinal ligament

**Which approximate derma-
tomal level is represented
by the:**

    Xiphoid process?   T7

    Umbilicus?   T10

    Inguinal ligament?   L1

What is the linea alba?   The midline fascial band that extends
from the symphysis pubis to the
umbilicus

**What is the linea semi-
lunaris?**   The curved groove just lateral to the
rectus abdominis muscle

What is the arcuate line?   A horizontal line midway between the
symphysis pubis and the umbilicus that
delineates the transition from the
aponeurotic posterior wall of the
rectus sheath to the transversalis
fascia

**How does the rectus sheath
differ:**

    Above the arcuate line?   The aponeurosis of the external oblique
muscle contributes to the anterior layer
of the sheath, the aponeurosis of the
transversus abdominis muscle contributes to the posterior layer, and the
aponeurosis of the internal oblique
muscle contributes to both layers

    Below the arcuate line?   The aponeuroses of all three flat muscles
form the anterior layer of the rectus
sheath, and the rectus abdominis muscle
lies in direct contact with the transversalis
fascia posteriorly.

Identify the borders of the inguinal triangle (Hesselbach's triangle) on the following figure:

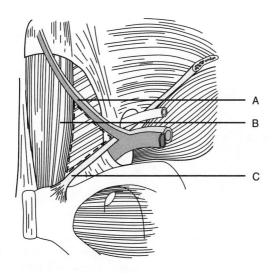

A = Inferior epigastric artery and vein
B = Lateral border of the rectus abdominis muscle (i.e., the linea semilunaris)
C = Inguinal ligament

The intercostal nerves run between which layers of the abdominal wall?

The intercostal nerves run between the internal oblique and transversus abdominis muscles, in the so-called "neurovascular plane."

What are the major nerves of the anterior abdominal wall?

The inferior six thoracic nerves (T7–T11) and the subcostal nerve (T12)

Which muscles are innervated by the subcostal nerve?

The external oblique, internal oblique, transversus abdominis, rectus abdominis, and pyramidalis muscles

Describe the lymphatic drainage of the anterior abdominal wall.

Above the umbilicus, drainage is to the axillary nodes. Below the umbilicus, drainage is to the superficial inguinal, external iliac, and aortic (lumbar) nodes.

POSTERIOR ABDOMINAL WALL

Identify the anterolateral and posterior abdominal wall muscles on the following figure:

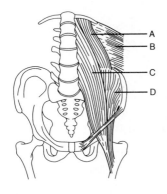

A = Quadratus lumborum muscle
B = Transversus abdominis muscle
C = Psoas major muscle
D = Iliacus muscle

What is the origin and insertion of the psoas major muscle?	The psoas major muscle runs from the transverse processes of the lumbar vertebrae and the bodies of vertebrae T12–L5 to the lesser trochanter of the femur.
Where does the iliacus muscle insert?	On the lesser trochanter of the femur (along with the psoas major muscle)
The medial arcuate ligament of the diaphragm is formed by a thickening of which fascia superiorly?	The iliac fascia
The lateral arcuate ligament of the diaphragm is formed by a thickening of which fascia superiorly?	The fascia of the quadratus lumborum muscle
The lumbar plexus lies within which muscle?	The psoas major muscle
What nerves contribute to the lumbar plexus?	The L1–L3 ventral primary rami and the superior branch of spinal nerve L4
What are the two largest branches of the lumbar plexus?	The femoral nerve and the obturator nerve

What are the four other branches of the lumbar plexus?	1. Ilioinguinal nerve 2. Iliohypogastric nerve 3. Genitofemoral nerve 4. Lateral femoral cutaneous nerve
Describe the course of the femoral nerve.	The femoral nerve arises from the lateral border of the psoas major muscle, descends in the groove between the psoas major and iliacus muscles, and enters the femoral triangle deep to the inguinal ligament (lateral to the femoral vein).
Which muscles are supplied by the femoral nerve?	The knee extensors
Describe the course of the obturator nerve.	The obturator nerve descends through the psoas major muscle and pierces the fascia to pass lateral to the internal iliac vessels and ureter. The obturator nerve leaves the pelvis and enters the lateral thigh via the obturator foramen.
Which muscles are supplied by the obturator nerve?	The thigh adductors
Which two nerves arise from the L1 segment of the lumbar plexus?	The ilioinguinal and iliohypogastric nerves
Which structures are innervated by the:	
Ilioinguinal nerve?	The abdominal wall muscles and the skin of the groin and genitalia
Iliohypogastric nerve?	The abdominal wall muscles and the skin of the hypogastric and gluteal regions
Which nerve from the lumbar plexus accompanies the spermatic cord (in men) or the round ligament of the uterus (in women) through the superficial inguinal ring?	The ilioinguinal nerve
Which nerve arises from the L1 and L2 segments of the lumbar plexus?	The genitofemoral nerve

Identify the lettered structures on the following illustration of the posterior abdominal wall:

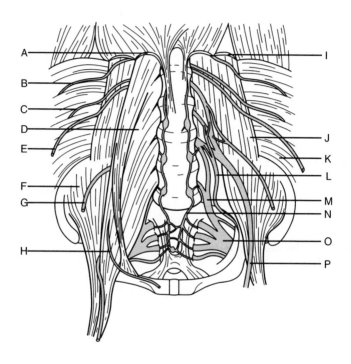

A = Medial arcuate ligament
B = Subcostal nerve
C = Iliohypogastric nerve
D = Psoas major muscle
E = Ilioinguinal nerve
F = Iliacus muscle
G = Lateral femoral cutaneous nerve
H = Genitofemoral nerve
I = Lateral arcuate ligament
J = Quadratus lumborum muscle
K = Transversus abdominis muscle
L = Obturator nerve
M = Lumbosacral trunk
N = Sympathetic trunk
O = Sciatic nerve
P = Femoral nerve

INGUINAL REGION

The inguinal ligament (Poupart's ligament) extends between which two structures?	The anterior superior iliac spine and the pubic tubercle
What forms the inguinal ligament?	The folded lower margin of the aponeurosis of the external oblique muscle

Which structures form the:

Superficial (external) inguinal ring?	The aponeurosis of the external oblique muscle
Deep (internal) inguinal ring?	The transversalis fascia

What forms the walls of the inguinal canal:

Anteriorly?	The aponeuroses of the external and internal oblique muscles
Posteriorly?	The transversalis fascia
Superiorly?	The arching fibers of the internal oblique and transversus abdominis muscles
Inferiorly?	The inguinal ligament

Which major structures pass through the inguinal canal in:

Men?	The spermatic cord and the ilioinguinal nerve
Women?	The round ligament of the uterus and the ilioinguinal nerve
What structures comprise the spermatic cord?	1. The ductus deferens (vas deferens) 2. The arteries of the ductus deferens (two) 3. The testicular artery 4. The testicular vein (continuous with the pampiniform plexus)

5. The genital branch of the
   genitofemoral nerve
6. Autonomic nerves
7. Lymphatic vessels draining the testes

**Which three structures form
the covering of the sper-
matic cord?**

1. The internal spermatic fascia
2. The cremaster muscle and cremasteric
   fascia
3. The external spermatic fascia

PERITONEUM

**What is the difference
between parietal and vis-
ceral peritoneum?**

The parietal peritoneum lines the inner
abdominal wall and diaphragm. The
visceral peritoneum covers the intra-
abdominal organs.

**What is the peritoneal
cavity?**

The potential space between the two
layers of peritoneum

**Is the peritoneal cavity
normally a closed space?**

Only in men; in women, the peritoneal
cavity communicates with the
reproductive tract through the
infundibulum of the fallopian tubes (this
explains how a pelvic infection can ascend
into the abdominal cavity in women).

**What is a peritoneal
ligament?**

A doubled fold of peritoneum that
extends between two organs

What is a peritoneal fold?

Peritoneum that is reflected away from
the abdominal wall by underlying blood
vessels or obliterated fetal vessels

What is a mesentery?

A doubled fold of peritoneum that
connects an abdominal organ to the
abdominal wall; vessels and nerves that
supply the organ are housed in between
the doubled layers of peritoneum

**What is an "intraperitoneal"
organ?**

An organ with a mesentery

**What is a "retroperitoneal"
organ?**

An organ that sits directly on the
posterior abdominal wall and is covered
anteriorly with visceral peritoneum

**Which gastrointestinal
structures are retroperi-
toneal?**

1. The pancreas
2. Most of the duodenum
3. The ascending and descending colon
Note that the appendix is **not**
retroperitoneal in most patients.

Name the mesenteries of the following organs:

Small intestine

Mesentery proper or "mesentery of the small intestine"

Transverse colon

Transverse mesocolon

Sigmoid colon

Sigmoid mesocolon

Appendix

Mesoappendix

Stomach

Greater omentum and lesser omentum

List six structures that are crossed by the root of the mesentery proper.

1. Duodenum
2. Aorta
3. Inferior vena cava
4. Psoas major muscle
5. Right ureter
6. Right testicular or ovarian vessels

What are the five sites of attachment of the greater omentum?

1. The greater curvature of the stomach
2. The first portion of the duodenum
3. The diaphragm
4. The spleen
5. The transverse colon

What are the three sites of attachment of the lesser omentum?

1. The lesser curvature of the stomach
2. The first portion of the duodenum
3. The liver

Which vessels run between the two layers of the lesser omentum?

The left and right gastric vessels

What is the name of the space posterior to the stomach and lesser omentum?

The omental bursa, or lesser peritoneal sac

What is the site where the omental bursa opens into the rest of the peritoneal cavity (i.e., the greater peritoneal sac) called?

The epiploic foramen (of Winslow)

What is the name of the right free margin of the lesser omentum?

The hepatoduodenal ligament (named aptly for its two attachments)

Which three important structures lie in the hepatoduodenal ligament?	1. The common hepatic artery 2. The portal vein 3. The common bile duct
What is the orientation of the hepatic artery, portal vein, and bile duct within the hepatoduodenal ligament?	The **P**ortal vein is **P**osterior, the hepatic artery is left anterior, and the bile duct is right anterior (think **LARD: L**eft **A**rtery, **R**ight **D**uct)
Name the double layer of parietal peritoneum that connects the liver to the diaphragm and the anterior abdominal wall.	The falciform ligament
Which blood vessels lie in the falciform ligament?	The paraumbilical veins
Which structure lies in the free margin of the falciform ligament?	The ligamentum teres (fibrosed)
What is the embryologic origin of the ligamentum teres?	The fetal left umbilical vein
Which structures form the:	
Median umbilical folds?	A remnant of the urachus
Medial umbilical folds?	Obliterated fetal umbilical arteries
Lateral umbilical folds?	Inferior epigastric vessels
What is the most posterior recess within the abdominal cavity?	The hepatorenal recess (because of its location, the hepatorenal recess is a common site for intra-abdominal fluid collections and abscesses)
Where are the:	
Paracolic gutters?	To the right and left of the ascending colon (referred to as the "right lateral paracolic gutter" and the "left lateral paracolic gutter," respectively)
Paravertebral gutters?	On either side of the vertebral column

ABDOMINAL VASCULATURE

ARTERIES

Label the branches of the abdominal aorta on the following figure:

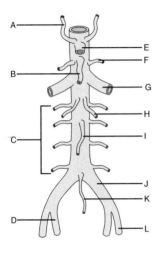

A = Inferior phrenic artery
B = Superior mesenteric artery
C = Lumbar arteries
D = Internal iliac artery
E = Celiac trunk
F = Middle suprarenal artery
G = Renal artery
H = Testicular or ovarian artery
I = Inferior mesenteric artery
J = Common iliac artery
K = Middle (median) sacral artery
L = External iliac artery

Which external anatomic landmark lies at the level of the abdominal aortic bifurcation?

The umbilicus

Where does the external iliac artery become the femoral artery?

At the inguinal ligament

Which three arteries arise from the celiac trunk?

1. Splenic artery
2. Left gastric artery
3. Common hepatic artery

Name the six major branches of the superior mesenteric artery.

1. Inferior pancreaticoduodenal artery
2. Middle colic artery
3. Ileocolic artery
4. Right colic artery
5. Jejunal arteries
6. Ileal arteries

Label the branches of the superior mesenteric artery on the following figure:

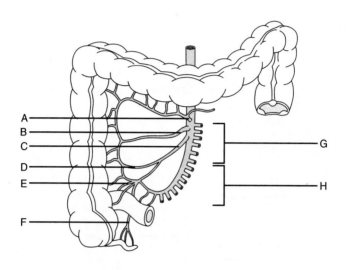

A = Middle colic artery
B = Right colic artery
C = Ileocolic artery
D = Ascending colic branches
E = Cecal branches
F = Appendiceal artery
G = Jejunal arteries
H = Ileal arteries

Name the four major branches of the ileocolic artery.

1. Ascending colic branches
2. Cecal branches
3. Appendiceal artery
4. Ileal and jejunal branches

Label the three major branches of the inferior mesenteric artery on the following figure:

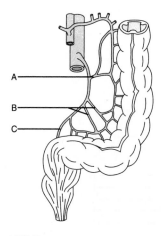

A = Left colic artery
B = Sigmoid arteries (two or three are usually present)
C = Superior rectal artery

VEINS

Label the tributaries of the abdominal inferior vena cava on the following figure:

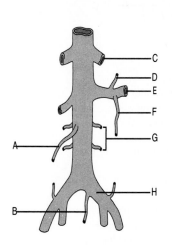

A = Right testicular or ovarian vein
B = Middle sacral vein
C = Hepatic vein
D = Inferior adrenal (suprarenal) vein
E = Left renal vein
F = Left testicular or ovarian vein
G = Lumbar veins
H = Left common iliac vein

What is the drainage of the:

Left testicular or ovarian vein?   Into the left renal vein

Right testicular or ovarian vein?   Directly into the inferior vena cava

ABDOMINAL VISCERA

Label the following diagram of the abdominal organs:

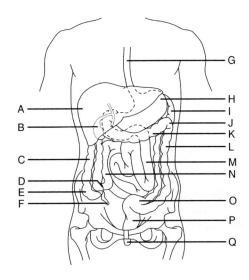

A = Liver
B = Gallbladder
C = Ascending colon
D = Ileocecal valve
E = Cecum
F = Vermiform appendix
G = Esophagus
H = Stomach
I = Spleen
J = Left colic (splenic) flexure
K = Transverse colon
L = Descending colon
M = Jejunum
N = Ileum
O = Sigmoid colon
P = Rectum
Q = Anus

Describe the extent of the:

Foregut?	From the oropharynx to the hepatopancreatic ampulla (of Vater)
Midgut?	From the hepatopancreatic ampulla (of Vater) to the distal third of the transverse colon
Hindgut?	From the distal transverse colon to the anus

What is the major arterial supply of the:

Foregut?	The celiac trunk
Midgut?	Superior mesenteric artery
Hindgut?	Inferior mesenteric artery

ESOPHAGUS

Which structure accompanies the esophagus through the diaphragm?

The vagus nerve

Is the left vagus nerve anterior or posterior?

Anterior (remember **LARP: L**eft **A**nterior, **R**ight **P**osterior)

Which arteries supply the abdominal esophagus?

The inferior phrenic artery and the left gastric artery

What is the venous drainage of the abdominal esophagus?

The azygos vein and the left gastric vein

Name the layers of the esophageal wall (from deep to superficial).

1. Mucosa
2. Submucosa
3. Muscularis
Note that unlike the rest of upper gastrointestinal tract, the esophagus has no serosa; because of this, esophageal cancer tends to invade outside structures early in its course, and accordingly is associated with a high mortality rate.

How does the muscular layer of the esophagus differ along its length?

The upper third of the esophagus has striated (voluntary) muscle, the lower third has smooth (involuntary) muscle, and the middle third has both striated and smooth muscle.

Which muscle functions as the upper esophageal sphincter?

The cricopharyngeus muscle

Where are the four most likely points of constriction or obstruction along the esophagus?

1. Upper esophageal sphincter
2. Aortic arch
3. Left main stem bronchus
4. Lower esophageal sphincter

The esophagus enters which part of the stomach?

The cardia

STOMACH

Identify the labeled structures on the following diagram of the stomach:

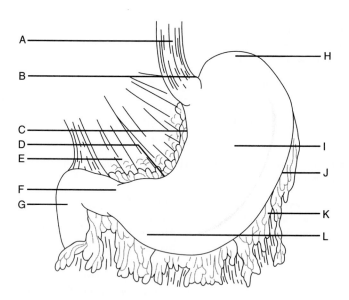

A = Esophagus
B = Cardiac notch
C = Lesser curvature of the stomach
D = Angular notch
E = Lesser omentum
F = Pylorus

G = Duodenum
H = Fundus
I = Body of the stomach
J = Greater curvature of the stomach
K = Greater omentum
L = Antrum

What part of the stomach contacts the diaphragm?

The fundus

The anterior stomach contacts which three structures?

1. The anterior abdominal wall
2. The left lobe of the liver
3. The diaphragm

What is the name of the large, longitudinal folds in the mucous membrane of the stomach?

Rugae; these folds flatten as the stomach fills with food

What are the three muscular layers of the stomach wall, from deep to superficial?

1. Oblique fibers
2. Circular fibers
3. Longitudinal fibers

Which arteries constitute the stomach's arterial supply?

1. The left and right gastric arteries (branches of the celiac artery and the hepatic artery, respectively)
2. The left and right gastroepiploic arteries (branches of the splenic artery and the gastroduodenal artery, respectively)
3. The short gastric arteries (branches of the splenic artery)

Which vessels run along the stomach's:

Lesser curvature?

The right and left gastric arteries

Greater curvature?

The right and left gastroepiploic arteries

**Identify the labeled arteries
on the following figure of
the stomach:**

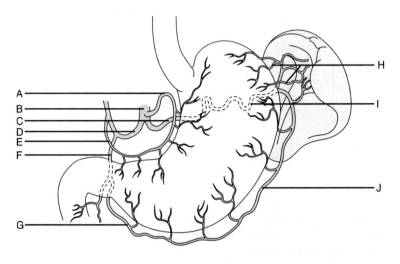

A = Left gastric artery
B = Celiac trunk
C = Splenic artery
D = Hepatic artery
E = Right gastric artery
F = Gastroduodenal artery
G = Right gastroepiploic artery
H = Short gastric arteries
I = Splenic artery
J = Left gastroepiploic artery

Are nerves on the anterior stomach derived from the left or right vagus?	The left (remember **LARP!**)
Describe the rotation of the stomach that takes place during fetal development.	The stomach rotates 90° clockwise; this is why the left vagus nerve is anterior, while the right vagus nerve is posterior.
Which portion of the stomach lies just proximal to the duodenum?	The pylorus
Which blood vessel marks the gastroduodenal junction and site of the pyloric orifice?	The prepyloric vein (of Mayo)

SMALL INTESTINE

How long is the adult small intestine?	Approximately 20 feet (6 meters)
What are plicae circulares (valvulae conniventes)?	Circular folds of small intestinal mucosa that are useful for differentiating the small bowel from the large bowel on radiographs (the plicae circulares are seen as complete circles)
Name the three parts of the small intestine.	Duodenum, jejunum, and ileum

Duodenum

What is the derivation of the name "duodenum?"	*Duodeni* is Latin for "twelve;" the duodenum is about 12 finger breadths (25 cm) long.
Is the duodenum intraperitoneal or retroperitoneal?	The proximal duodenum (i.e., the first 2.5 cm or so) has a mesentery and is therefore intraperitoneal; the remainder of the duodenum is retroperitoneal.
Which arteries supply the duodenum?	1. The superior pancreaticoduodenal arteries (branches of the gastroduodenal artery) 2. The inferior pancreaticoduodenal arteries (branches from the superior mesenteric artery) These vessels anastomose to form arterial arcades.
What are the four parts of the duodenum?	1. Superior 2. Descending 3. Inferior (horizontal) 4. Ascending
What is the approximate vertebral level of the:	
First portion of the duodenum?	Approximately L1
Third portion of the duodenum?	Approximately L3

What is another name for the beginning of the first portion of the duodenum?	The duodenal bulb (ampulla)
Which artery lies directly behind the first portion of the duodenum?	The gastroduodenal artery (the location of this major artery makes it susceptible to invasion by penetrating posterior ulcers in the first portion of the duodenum, which can lead to severe gastrointestinal hemorrhage)
Where do the common bile duct and the pancreatic ducts enter the duodenum?	On the posteromedial aspect of the second part of the duodenum
What is the name of the dilatation formed by the junction of the common bile duct and the main pancreatic duct (of Wirsung) just proximal to their opening into the duodenum?	The hepatopancreatic ampulla (of Vater)
What is the name of the muscular sphincter at the hepatopancreatic ampulla (of Vater)?	The sphincter of Oddi
What does the sphincter of Oddi do?	It regulates the entry of bile and pancreatic enzymes into the small intestine.
What marks the spot on the duodenal wall where the hepatopancreatic ampulla (of Vater) enters the intestinal lumen?	The major duodenal papilla
What is the minor duodenal papilla?	The minor duodenal papilla is a communication between the accessory pancreatic duct (of Santorini) and the duodenum. When present (10% of patients), the minor duodenal papilla usually lies several centimeters above the opening of the main pancreatic duct (of Wirsung). In the rest of the population, the accessory pancreatic duct merges with the main pancreatic duct before it enters duodenum.

Jejunum and ileum

What is the ligament of Treitz?	A well-marked peritoneal fold at the junction of the fourth part of the duodenum with the jejunum
Describe the path of blood flow from the aorta to the jejunum.	The aorta gives rise to the **superior mesenteric artery,** which runs in the root of the mesentery, giving off 15–18 **jejunal** and **ileal branches** that run between the two layers of mesentery and unite to form loops called **arterial arcades.** These arcades then form **straight vessels (vasa recta),** which pass alternately to opposite sides of the intestine and enter the intestine on the mesenteric border.
How does the jejunum differ in appearance from the ileum?	1. Thicker wall 2. Larger diameter 3. More vascular (redder) 4. Larger and more developed plicae circulares 5. Longer vasa rectae 6. Less mesenteric fat 7. Lacks Peyer's patches (lymphoid aggregates along the anti-mesenteric border of the ileum)
Where are the mesenteric attachments of the:	
Jejunum?	Above and to the left of the aorta
Ileum?	Below and to the right of the aorta (accordingly, the jejunum tends to occupy more of the left upper quadrant and the ileum tends to occupy more of the right lower quadrant)
Which structure marks the end of the small intestine?	The ileocecal valve

LARGE INTESTINE

How long is the adult large intestine?	Approximately 1.5 meters

Identify the labeled structures on the following diagram of the large intestine:

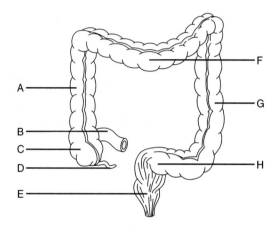

A = Ascending colon
B = Terminal ileum
C = Cecum
D = Vermiform appendix
E = Rectum
F = Transverse colon
G = Descending colon
H = Sigmoid colon

Name the seven parts of the large intestine, from proximal to distal.

1. Vermiform appendix
2. Cecum
3. Ascending colon
4. Transverse colon
5. Descending colon
6. Sigmoid colon
7. Rectum

Name three features that differentiate the large intestine from the small intestine on gross inspection.

1. Teniae coli
2. Haustra
3. Appendices epiploicae

What are the teniae coli?

The three thin bands of muscle that run longitudinally along the entire length of the ascending, transverse, and descending colon

Where do the three taeniae coli converge?

At the appendix (this can help you locate the appendix during surgery)

What are haustra, and what causes them?

Characteristic sacculations of the large intestine which are created by contraction of the taeniae coli; on radiographs, the haustra appear as incomplete circles

What are appendices epiploica (epiploic appendages)?

The small, fat-filled peritoneal sacs along the taeniae coli

Which parts of the large intestine are retroperitoneal?

The ascending and descending colon

What portion of the large intestine has the widest diameter?

The cecum (typically measures 7–9 cm in diameter)

What arteries supply the cecum?

The anterior and posterior cecal arteries (branches of the ileocolic artery, which is a branch of the superior mesenteric artery)

Where is the vermiform appendix located?

Its position is variable; it is often found in the pelvis or behind the cecum in the retrocecal fossa

What artery supplies the vermiform appendix?

The appendicular artery, a branch of the ileocolic artery

Does the appendix have a mesentery?

Yes, the appendix has a mesentery even though the cecum does not. The mesoappendix suspends the appendix from the mesentery of the terminal ileum.

What arteries supply the ascending colon?

The ileocolic and right colic arteries (both branches of the superior mesenteric artery)

Where is the point of transition between the superior and inferior mesenteric arterial blood supply to the large intestine?

At the beginning of the distal third of the transverse colon (the site of embryologic transition from midgut to hindgut)

What are the clinical implications associated with this transitional area?

This area, which is between blood supplies, is especially prone to ischemia during times of decreased blood flow to the colon (the so-called "watershed" effect).

Describe the mesentery of the transverse colon.	The transverse mesocolon is a double layer of peritoneum that suspends the transverse colon from the posterior abdominal wall and connects the transverse colon to the pancreas and greater omentum.
Which arteries supply the descending colon?	The left colic and superior sigmoid arteries (both branches of the inferior mesenteric artery)
How can one grossly identify the beginning of the sigmoid colon?	The ends of the teniae coli mark the end of the descending colon and the beginning of the sigmoid colon; in addition, the descending colon is retroperitoneal and the sigmoid colon is intraperitoneal.
Which arteries supply the sigmoid colon?	The sigmoid arteries
How does the rectum differ from the sigmoid colon?	1. The rectum lacks a peritoneal covering distally. 2. The teniae coli broaden to form a complete longitudinal layer. 3. The rectum has no haustra or appendices epiploicae, but it does have transverse rectal folds (which the sigmoid colon lacks).

PANCREAS

Identify the labeled structures on this figure of the pancreas and surrounding organs:

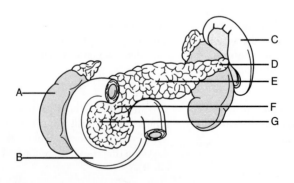

A = Right kidney
B = Duodenum
C = Spleen
D = The tail of the pancreas
E = The body of the pancreas
F = The neck of the pancreas
G = The head of the pancreas

The head of the pancreas lies in intimate contact with which organ? How about the tail?

The head of the pancreas lies in the duodenum, and the tail contacts the spleen (the pancreas is "cradled in the arms of the duodenum and tickles the spleen with its tail").

What is the uncinate process?

A hook-shaped projection from the lower aspect of the head of the pancreas that extends superiorly and to the left and lies between the superior mesenteric vessels and the aorta

Which major vessels course along the pancreas?

The splenic artery (which originates at the celiac trunk) and the splenic vein (pancreatitis can lead to splenic vein thrombosis)

Which arteries supply the head of the pancreas?

1. The anterior superior and posterior superior pancreaticoduodenal arteries (branches of the gastroduodenal artery)
2. The anterior inferior and posterior inferior pancreaticoduodenal arteries (branches of the superior mesenteric artery)

What arteries supply the body and tail of the pancreas?

1. The splenic artery
2. The dorsal pancreatic artery
3. The great pancreatic artery (pancreatica magna)
4. The caudal pancreatic artery

**Identify the labeled arteries
on the following figure of
the pancreas:**

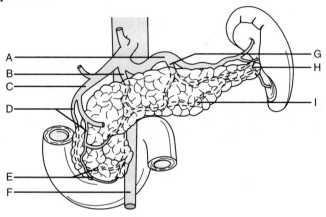

A = Splenic artery
B = Dorsal pancreatic artery
C = Gastroduodenal artery
D = Posterior and anterior superior
        pancreaticoduodenal arteries
E = Posterior and anterior inferior
        pancreaticoduodenal arteries
F = Superior mesenteric artery
G = Great pancreatic artery
H = Caudal pancreatic arteries
I = Inferior pancreatic artery

What is the duct of Wirsung? Of Santorini?	The main pancreatic duct and the accessory pancreatic duct, respectively
The main pancreatic duct (of Wirsung) normally joins which structure?	The common bile duct

LIVER

Which structure traditionally divides the liver into right and left lobes, anatomically speaking?	The falciform ligament
What line divides the liver into functionally independent right and left lobes?	Cantlie's line (an imaginary line drawn across the diaphragmatic surface of the liver that runs from the gallbladder to the inferior vena cava)

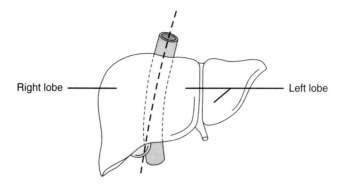

Right lobe ——————————— ——————— Left lobe

What does "functionally independent" mean?

The right and left functional lobes have their own arterial and portal blood supply, venous drainage, and biliary drainage.

What are the four lobes of the liver?

1. Right lobe
2. Left lobe
3. Quadrate lobe
4. Caudate

From a functional standpoint, the caudate and quadrate lobes are part of which side of the liver?

The left

What is an alternative scheme for dividing the liver anatomically and functionally?

The French (hepatic segment) system, based on the branching of the portal vein and hepatic artery

Identify the liver segments according to the French system on the figure below:

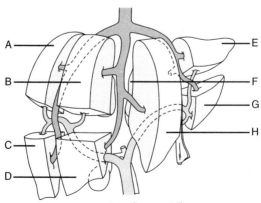

A = Segment 7
B = Segment 8

C = Segment 6
D = Segment 5
E = Segment 2
F = Segment 1
G = Segment 3
H = Segment 4
An easy way to remember the order of the segments: start by identifying segment 1, then move in a clockwise direction.

What is the name of the space between the liver and the diaphragm?

The subphrenic recess

What is the "bare area" of the liver?

The "bare area" of the liver is the triangular area of the liver that is **not covered by peritoneum.** The two layers of the falciform ligament separate superiorly and are reflected onto the diaphragm as the coronary ligament, leaving this part of the liver in direct contact with the diaphragm without any peritoneal covering.

What is the porta hepatis?

A transverse fissure on the visceral surface of liver; the peritoneum opens at this fissure to transmit the portal vein, hepatic artery proper, and the right and left hepatic bile ducts.

Name the four organs that contact the visceral surface of the liver.

1. Stomach
2. Duodenum
3. Gallbladder
4. Colon (at the right colic flexure)

How is the liver attached to the stomach and duodenum?

By the lesser omentum, which has two parts, the hepatogastric ligament and the hepatoduodenal ligament

Identify the labeled structures on the following views of the liver and associated area:

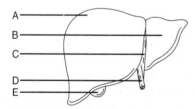

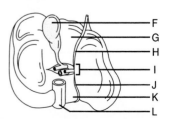

A = Right liver lobe
B = Left liver lobe
C = Falciform ligament
D = Ligamentum teres
E = Gallbladder
F = Gallbladder
G = Quadrate liver lobe
H = Ligamentum teres
I = Porta hepatis
J = Ligamentum venosum
K = Caudate liver lobe
L = Inferior vena cava

Describe the path of arterial blood flow from the aorta to the liver.	The common hepatic artery arises from the celiac trunk and bifurcates to form the gastroduodenal artery and the hepatic artery proper. The hepatic artery proper divides into right and left branches shortly before entering the liver at the porta hepatis.
What percentage of blood flow to the liver is provided by the:	
Hepatic artery?	Approximately 30%
Portal vein?	The remaining 70%
What is the general function of the portal vein?	The portal vein collects blood from the gastrointestinal tract, gallbladder, pancreas, and spleen, and carries it to the liver.
Which two veins unite to form the portal vein?	The superior mesenteric vein and the splenic vein
What happens to the inferior mesenteric vein?	The inferior mesenteric vein joins the splenic vein before the latter unites with the superior mesenteric vein to form the portal vein.
Name the four communications between the portal circulation and the systemic circulation that provide collateral portal circulation in the event of obstruction in the liver or portal vein.	1. The left gastric (coronary) vein anastomoses with the esophageal veins of the azygos system. 2. The superior rectal vein anastomoses with the middle and inferior rectal veins. 3. The paraumbilical veins anastomose with the epigastric veins of the anterior abdominal wall. 4. Tributaries of the splenic and

pancreatic veins anastomose with the left renal vein.

In patients with portal hypertension, blood tends to be diverted into the systemic circulation at the sites of the portosystemic anastomoses. What clinical conditions may be seen when this occurs?

Esophageal varices (site #1), hemorrhoids (site #2), and caput medusae (site #3)—think "gut, butt, and caput" plus the retroperitoneum (site #4)

What is the venous drainage of the liver?

The hepatic veins drain directly into the inferior vena cava, just inferior to the diaphragm

What is the primary lym-phphatic drainage of the liver?

The lymphatics drain to the hepatic nodes, which are located alongside the hepatic vessels in the lesser omentum. The hepatic nodes then drain to the celiac nodes.

GALLBLADDER AND BILE DUCTS

What is the approximate capacity of the adult gall-bladder?

30–50 ml

Label the parts of the gall-bladder on the following figure:

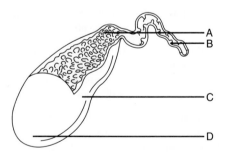

A = Infundibulum of the gallbladder (Hartman's pouch)
B = Cystic duct
C = Body of the gallbladder
D = Fundus of the gallbladder

What are the folds in the mucous membrane of the cystic duct called?

The spiral valves of Heister (impacted gallstones often lodge in the spiral valves)

Which artery supplies the gallbladder?

The cystic artery, which is normally a branch of the right hepatic artery; however, there is tremendous variability in the origin of the cystic artery and other vessels near the porta hepatis

What is the venous drainage of the gallbladder?

Most of the gallbladder fundus and body drains directly into the liver.

What is the gallbladder's lymphatic drainage?

The lymphatics drain to the hepatic lymph nodes, which in turn drain to the celiac nodes.

Define Calot's triangle.

It's the anatomic triangle bordered by the liver, the cystic duct, and the hepatic duct.

Which key structure lies within Calot's triangle?

The cystic artery

Trace the flow of bile from the gallbladder to the intestine.

Bile flows out of the gallbladder infundibulum and into the cystic duct. The cystic duct joins the common hepatic duct to form the common bile duct, which travels with the hepatic artery and portal vein in the lower free edge of the lesser omentum. After passing posterior to the superior portion of the duodenum, the common bile duct is usually joined by the main pancreatic duct (of Wirsung); it then enters the hepatopancreatic ampulla (of Vater). The hepatopancreatic ampulla (of Vater) opens into the second portion of the duodenum at the major duodenal papilla, which is about 8–10 centimeters distal to the pylorus.

What is the narrowest point in the biliary system?

The hepatopancreatic ampulla (of Vater)

Name the two muscular sphincters in the distal biliary tract.

1. The choledochal sphincter (located at the distal end of the bile duct
2. The hepatopancreatic sphincter (of Oddi), located at the hepatopancreatic ampulla (of Vater)

SPLEEN

How big is a normal adult spleen?	About the size of a fist
Which four abdominal organs are normally in contact with the spleen?	1. The stomach 2. The left kidney 3. The colon 4. The pancreas
Which of these organs contacts the hilum of the spleen?	The pancreas
What other structure does the spleen lie in direct contact with?	The diaphragm
Which vessels run in the gastrolienal (gastrosplenic) ligament?	The short gastric and the left gastroepiploic vessels
Which vessels run in the lienorenal (splenorenal) ligament?	The splenic artery and vein
Which artery constitutes the spleen's primary blood supply?	The splenic artery, the largest branch from the celiac trunk
Describe the course of the splenic artery.	The splenic artery runs from the celiac trunk along the superior border of the pancreas in the lienorenal ligament. It then divides within the ligament into several branches, which enter the spleen at the hilum.
Describe the spleen's venous drainage.	Venous drainage of the spleen is via the splenic vein, which unites with the superior mesenteric vein posterior to the pancreas to form the portal vein.
What is an accessory spleen?	Heterotopic splenic tissue, usually located near the splenic hilum (seen in approximately 20% of the population)

KIDNEYS, ADRENAL GLANDS, AND URETERS

The kidneys normally lie at approximately what vertebral level?	At the level of the T12–L3 vertebrae
Are both kidneys normally at the same level?	No. Because of the large right lobe of the liver, the right kidney tends to be located more inferiorly than the left.
Are the kidneys retroperitoneal or intraperitoneal?	Retroperitoneal
The kidneys lie in close association with which muscle?	The psoas major muscle
What is the name of the capsule that surrounds the kidney?	The renal fascia (Gerota's fascia)
What lies between the renal fascia and the peritoneum of the posterior abdominal wall?	Pararenal fat
What lies between the renal fascia and the kidney?	Perirenal fat
What is the name of the kidney's medial concave margin?	The renal sinus (the site of the renal hilum and the renal pelvis)
Describe the gross organization of the kidney on cut section.	The renal parenchyma is organized into a cortex and medulla; the cortex is the portion of the parenchyma closest to the renal fascia.
What arteries supply the kidneys?	The renal arteries, which are branches of the abdominal aorta at the level of the L1–L2 vertebrae
Name the three structures that enter or leave the kidney at its hilum.	1. Renal vein 2. Renal artery 3. Renal pelvis
What is the orientation of these structures?	The renal vein is anterior, the renal artery is posterior, and the renal pelvis is posterior to the artery.

What is the renal pelvis?

The area of transition between the kidney and the ureter

Trace the flow of urine from the nephron to the bladder.

Urine flows from the collecting ductules of the nephron into the minor calices. The minor calices empty into the major calices (wide, cup-shaped structures in the renal sinus). The major calices then empty into the renal pelvis, which is continuous with the ureter.

Identify the labeled structures on the following figure of the kidney and ureter:

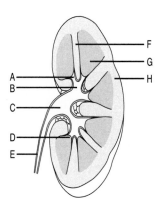

A = Minor calix
B = Major calix
C = Renal pelvis
D = Renal papilla
E = Ureter
F = Renal column
G = Renal pyramid
H = Renal cortex

Where are the adrenal glands located?

The adrenal glands lie on the anteromedial aspect of the superior pole of the kidneys, enveloped by the renal fascia.

Which arteries supply the adrenal gland?

1. The superior adrenal (suprarenal) artery, a branch of the inferior phrenic artery
2. The middle adrenal artery, a branch from the abdominal aorta
3. The inferior adrenal artery, a branch of the renal artery

What is the adrenal gland's venous drainage?

The suprarenal veins drain to the renal vein on the left, but directly to the inferior vena cava on the right.

POWER REVIEW

Which approximate derma-tomal level is represented by the:	
Xiphoid process?	T7
Umbilicus?	T10
Inguinal ligament?	L1
In which direction do the fibers of the external oblique muscle course?	Inferomedially
The internal oblique muscle?	Inferolaterally
The transversus abdominis muscle?	Transversely (horizontally)
What forms the inguinal ligament (Poupart's ligament)?	The folded lower margin of the aponeurosis of the external oblique muscle
The inguinal ligament runs between which two structures?	The anterior superior iliac spine and the pubic tubercle
Which major structures pass through the inguinal canal in:	
Men?	The spermatic cord and the ilioinguinal nerve
Women?	The round ligament of the uterus and the ilioinguinal nerve
Which structures in the abdominal cavity are retro-peritoneal?	1. The pancreas 2. Most of the duodenum 3. The ascending and descending colon 4. The aorta 5. The inferior vena cava 6. The kidneys 7. The ureters Note that the appendix is not retroperitoneal in most patients!

What structure lies in the free margin of the falciform ligament?	The ligamentum teres (the fetal left umbilical vein remnant)
What is the most posterior recess within the abdominal cavity?	The hepatorenal recess (Morrison's pouch)

What is the extent of the:

Foregut?	From the oropharynx to the hepatopancreatic ampulla (of Vater)
Midgut?	From the ampulla of Vater to the distal transverse colon
Hindgut?	From the distal transverse colon to the anus

What artery supplies the:

Foregut?	Celiac artery
Midgut?	Superior mesenteric artery
Hindgut?	Inferior mesenteric artery
Which two veins unite to form the portal vein?	The superior mesenteric vein and the splenic vein
Is the left vagus nerve anterior or posterior?	Anterior (remember **LARP**)
How does the muscular layer of the esophagus differ along its length?	The upper third is striated muscle, the lower third is smooth muscle, and the middle third is mixed.
What are the major structural differences between the small and large intestine?	The small intestine has plicae circulares, whereas the large intestine has haustra, taenia coli, and appendices epiploicae.
Which major vessels run with the pancreas?	The splenic artery and vein
Which structures enter the porta hepatis?	The hepatic artery, portal vein, and bile ducts
The cystic artery normally branches from which artery?	The right hepatic artery
Which kidney lies higher?	The left is higher than the right.

10 The Pelvis and Perineum

BONES AND LIGAMENTS

Which five bones make up the pelvis?

1. Ilium
2. Ischium
3. Pubis
4. Sacrum
5. Coccyx

Identify the lettered components of the hip bone on the following illustration:

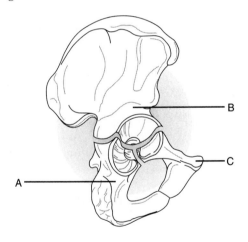

A = Ischium
B = Ilium
C = Pubis

Describe the surface anatomy of the ilium.

An upper ala (wing) forms the iliac crest.

The most superior point of the iliac crest as palpated posteriorly is located at which vertebral level?

L4

What is the name of the anterior and posterior terminations of the iliac crest?	The anterior superior iliac spine and the posterior superior iliac spine
Which anterior thigh muscle originates from the inner surface of the ilium (iliac fossa)?	The iliacus muscle
Name the sharp bony landmark on the medial side of the ischium.	Ischial spine
What structure attaches to the ischial spine?	The sacrospinous ligament
What are the articulations of the pubis?	The superior ramus articulates with the ilium at the iliopubic eminence; the inferior ramus articulates with the ischial ramus to form the pubic arch.
What is the acetabulum?	A deep hemispherical cup on the lateral surface of the pelvis where the head of the femur articulates with the pelvis
Which bones of the pelvis contribute to the acetabulum?	The ilium, ischium, and pubis
What bony foramen lies between the body of the ischium and the rami of the pubis bones?	The obturator foramen
What structures pass through the obturator foramen?	The obturator nerve, artery, and vein

Identify the labeled structures on the following illustration of the hip bone:

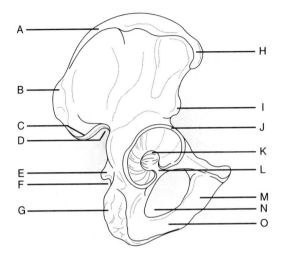

A = Iliac crest
B = Posterior superior iliac spine
C = Posterior inferior iliac spine
D = Greater sciatic notch
E = Ischial spine
F = Lesser sciatic notch
G = Ischial tuberosity
H = Anterior superior iliac spine
I = Anterior inferior iliac spine
J = Acetabular labrum
K = Acetabular fossa
L = Acetabular notch
M = Inferior pubic ramus
N = Obturator foramen
O = Ischial ramus

What are the four major articulations of the pelvis?

1. The lumbosacral joint
2. The sacroiliac joints
3. The sacrococcygeal joints
4. The symphysis pubis

Which of these are cartilaginous joints?

The symphysis pubis, the sacrococcygeal joint, and the lumbosacral joint

Which three ligaments support the sacroiliac joint?

1. The anterior sacroiliac ligament
2. The posterior sacroiliac ligament
3. The interosseus ligaments

Which two ligaments support the symphysis pubis?	The superior pubic ligament and the superior arcuate ligament
Which three ligaments support the sacrococcygeal joint?	The anterior, posterior, and lateral sacrococcygeal ligaments
Which ligament creates the greater and lesser sciatic foramina?	The sacrotuberous ligament, which extends from the sacrum to the ischial tuberosity

What structures pass out of the pelvic cavity through the greater sciatic foramen?

1. The piriformis muscle
2. The sciatic nerve
3. The internal pudendal artery and vein
4. The pudendal nerve
5. The superior and inferior gluteal vessels and nerves
6. The posterior femoral cutaneous nerve
7. Nerves to the quadratus femoris and obturator internus muscles

What structures pass into the pelvic cavity through the lesser sciatic foramen?

1. The internal pudendal artery and vein
2. The pudendal nerve
3. The nerve to the obturator internus muscle
4. The tendon of the obturator internus muscle

Identify the labeled structures on the following diagram of the bony pelvis:

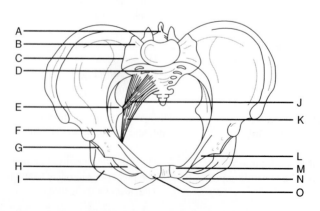

A = First sacral spinous process
B = Lateral mass of the sacrum
C = Sacroiliac joint
D = Sacral promontory
E = Ischial spine
F = Iliopectineal (arcuate) line
G = Acetabulum
H = Obturator foramen
I = Ischial ramus
J = Sacrospinous ligament
K = Sacrotuberous ligament
L = Superior ramus of the pubis
M = Pubic crest
N = Body of the pubis
O = Pubic tubercle

**Identify the labeled
structures on the following
medial view of the hip bone
and associated structures:**

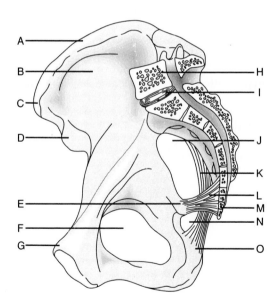

A = Iliac crest
B = Iliac fossa
C = Anterior superior iliac spine
D = Anterior inferior iliac spine
E = Ischial spine

F = Obturator foramen
G = Pubic tubercle
H = Vertebral body of L5
I = Lumbosacral joint
J = Greater sciatic foramen
K = Sacrotuberous ligament
L = Sacrospinous ligament
M = Coccyx
N = Lesser sciatic foramen
O = Ischial tuberosity

What is the pelvic inlet?

The superior rim of the pelvic cavity

What are the other names for the pelvic inlet?

Pelvic brim, superior pelvic aperture

What are the boundaries of the pelvic inlet:

Posteriorly?

The sacral promontory

Laterally?

The iliopectineal (arcuate) line

Anteriorly?

The superior margin of the symphysis pubis

What is the pelvic outlet?

The inferior rim of the pelvic cavity

What are the other names for the pelvic outlet?

The inferior pelvic aperture

What are the boundaries of the pelvic outlet:

Posteriorly?

The coccyx

Laterally?

The ischial tuberosities

Anteriorly?

The inferior margin of the symphysis pubis

What is the pelvis major (false pelvis)?

The wide portion of the bony pelvis superior to the pelvic brim (technically, the pelvis major is a part of the abdominal cavity)

What is the pelvis minor (true pelvis)?

The portion of the bony pelvis inferior to the pelvic brim and superior to the pelvic

outlet (technically, the pelvis minor
constitutes the pelvic cavity)

List five ways the pelvis differs between women and men.

1. In women, the bones of the pelvis are smaller, lighter, and thinner.
2. In women, the sacrum is broader and shorter.
3. In women, the suprapubic arch and the greater sciatic notch are wider.
4. In women, the pelvic inlet is ovoid, while in men, it is heart-shaped.
5. In women, the ischial tuberosities are everted, thereby enlarging the pelvic outlet.

The architecture of the pelvis can vary among individuals. Identify each of the four pelvis types on the figure below:

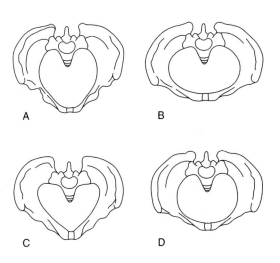

A

B

C

D

A = Anthropoid
B = Platypelloid
C = Android
D = Gynecoid

PELVIC MUSCLES AND FASCIAE

What structures form the:

    Anterior pelvic wall?

    The pubic bones, the symphysis pubis, the obturator internus muscle, and the obturator internus (parietal pelvic) fascia

    Lateral pelvic wall?

    The obturator internus muscle and its associated fascia, the ox coxae (below the pelvic brim), and the sacrotuberous and sacrospinous ligaments

    Posterior pelvic wall?

    The sacrum, coccyx, ligaments, and piriformis muscle

What structures form the floor of the pelvis?

The pelvic and urogenital diaphragms

Which muscles comprise the pelvic diaphragm?

The levator ani and coccygeus muscles

Identify the labeled structures on the following figure:

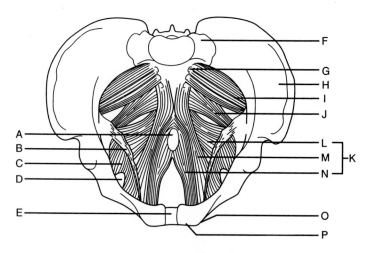

A = Anal canal
B = Tendinous arch of the obturator internus fascia
C = Obturator internus muscle

D = Obturator canal
E = Symphysis pubis
F = Sacrum
G = Sacral foramen
H = Iliac fossa
I = Piriformis muscle
J = Coccygeus muscle
K = Levator ani muscle group
L = Iliococcygeus muscle
M = Pubococcygeus muscle
N = Puborectalis muscle
O = Pubic tubercle
P = Pubic crest

Which muscles comprise the urogenital diaphragm?

1. The deep transverse perineal muscles, running transversely anteriorly and posteriorly
2. The sphincter urethrae muscle (and the sphincter vaginae muscle, in women)

The superior fascia of the urogenital diaphragm is continuous with which fascial layer?

The obturator internus (pelvic parietal) fascia

MUSCLES

Obturator internus muscle

Origin?

The pelvic surface of the obturator internus fascia and the surrounding parts of the ilium and pubis

Insertion?

The greater trochanter of the femur

Innervation?

The obturator nerve (from the sacral plexus)

Action?

Lateral rotation of the thigh when the hip joint is extended

What structure covers the obturator internus muscle?

The obturator internus fascia

What structure is formed by the obturator internus (pelvic parietal) fascia?

The pudendal canal, which transmits the pudendal nerve, artery, and vein

Piriformis muscle

Origin?	The anterolateral sacrum and the sacrotuberous ligament
Insertion?	The greater trochanter of the femur
Innervation?	The sacral plexus
Action?	Laterally rotates the thigh when the hip joint is extended and abducts the thigh when the hip joint is flexed

Levator ani muscle group

Which three muscles comprise the levator ani muscle group?

1. Puborectalis muscle
2. Pubococcygeus muscle
3. Iliococcygeus muscle

As a group, where do the levator ani muscles:

Originate?

From the body of the pubis, the tendinous arch of the obturator internus fascia, and the ischial spine

Insert?

On the perineal body (a fibromuscular mass anterior to the anus), the anococcygeal ligament (the median fibrous intersection of the pubococcygeus muscles), and the walls of the prostate (or vagina), rectum, and anal canal

What is the innervation of the levator ani muscle group?

Spinal nerves S3–S5 and the pudendal nerve

What are the actions of the levator ani muscle group?

1. Supports the pelvic viscera and counteracts increases in the intra-abdominal pressure
2. Elevates the floor of the pelvic cavity, assisting in compression of the contents of the pelvic and abdominal cavities

What is the action of the puborectalis muscle in particular?

The puborectalis muscle maintains the anorectal angle, which permits voluntary control of defecation

What is the anorectal angle?   The angle between the rectum and the anal canal; contraction of the puborectalis muscle holds the anorectal junction anteriorly, preventing the passage of feces from the rectum into the anal canal

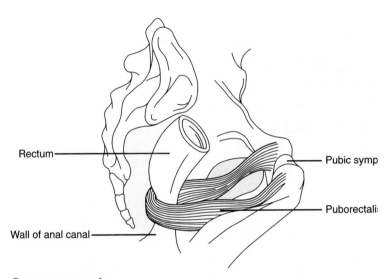

Rectum

Pubic symp

Puborectali:

Wall of anal canal

Coccygeus muscle

Origin?   The lateral pelvic surface of the ischial spine and the sacrospinous ligament

Insertion?   The medial and lateral margin of the coccyx and vertebra S5

Innervation?   Branches of spinal nerves S4 and S5

Action?   Assists the levator ani muscle group in supporting the pelvic viscera; supports and pulls the coccyx anteriorly

FASCIAE

The pelvic fascia is consistent with which fascial layer of the abdominal wall?   The transversalis fascia

What are the two layers of pelvic fascia?

1. Parietal pelvic fascia (covers the surfaces of the obturator internus, piriformis, levator ani, and coccygeus muscles, both superiorly and inferiorly)
2. Visceral pelvic fascia (binds pelvic organs to each other and to the parietal fascia)

What is name of the thickening of the obturator internus fascia from which the levator ani muscle group arises?

The tendinous arch

Identify the labeled structures on the following figure:

A = Bladder
B = Obturator internus muscle and fascia
C = Pudendal canal
D = Ischioanal (ischiorectal) fossa
E = Peritoneum
F = Pelvic diaphragm
G = Prostate gland
H = Superior fascia of the urogenital diaphragm

I = Urogenital diaphragm
J = Inferior fascia of the urogenital diaphragm (perineal membrane)
K = Bulb of the penis
L = Superficial perineal space

PELVIC VASCULATURE

ARTERIES

The common iliac artery divides into the external and the internal iliac arteries at which vertebral level?	L5–S1
What is the first branch of the internal iliac artery?	The iliolumbar artery
How does the internal iliac artery terminate?	By dividing into anterior and posterior branches
What are the three branches from the posterior internal iliac artery?	1. Iliolumbar artery 2. Lateral sacral artery 3. Superior gluteal artery
How does the superior gluteal artery leave the pelvis?	Through the superior part of the greater sciatic foramen (i.e., the part above the piriformis muscle)
What are the eight branches from the anterior internal iliac artery?	1. Umbilical artery 2. Superior vesical arteries 3. Uterine artery 4. Vaginal artery (in women) or inferior vesical artery (in men) 5. Middle rectal artery 6. Obturator artery 7. Internal pudendal artery 8. Inferior gluteal artery
How does the obturator artery leave the pelvis?	Through the obturator foramen
How does the inferior gluteal artery leave the pelvis?	Through the inferior part of the greater sciatic foramen

**Identify the labeled arteries
and associated structures
of the pelvic region on the
following figure:**

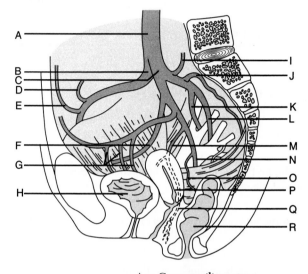

A = Common iliac artery
B = External iliac artery
C = Internal iliac artery
D = Deep circumflex iliac artery
E = Inferior epigastric artery
F = Obturator artery
G = Superior vesical arteries
H = Bladder
I = Iliolumbar artery
J = Lateral sacral artery
K = Superior gluteal artery
L = Inferior gluteal artery
M = Uterine artery
N = Internal pudendal artery
O = Middle rectal artery
P = Uterus
Q = Vaginal artery
R = Rectum

**What are the five branches
of the internal pudendal
artery?**

1. Inferior rectal artery
2. Perineal artery
3. Superficial perineal artery
4. Deep artery of the penis or clitoris
5. Dorsal artery of the penis or clitoris

What are the two terminal branches of the internal pudendal artery?	The deep and dorsal arteries of the penis (in men) or the clitoris (in women)
Where do the ovarian arteries arise?	Anteriorly from the abdominal aorta, just inferior to the renal arteries
Where do the testicular arteries arise?	Anteriorly from the abdominal aorta, just inferior to the renal arteries

LYMPHATICS

What is the drainage of the pelvic lymph nodes (i.e., the external iliac and internal iliac nodes)?	The common iliac nodes
What is the drainage of the perineal lymph nodes?	The perineal lymph nodes drain to the external pudendal nodes, which in turn drain to the superficial inguinal nodes.

PELVIC INNERVATION

SOMATIC INNERVATION

Which two major nerves are derived from the sacral plexus?	The sciatic nerve and the pudendal nerve
The pudendal nerve follows the course of which artery?	The internal pudendal artery
What are the two distal branches of the pudendal nerve?	1. The perineal branches 2. The dorsal nerve of the penis (in men) or clitoris (in women)

Identify the branches of the pudendal artery on the right side of the figure, and the branches of the pudendal nerve on the left:

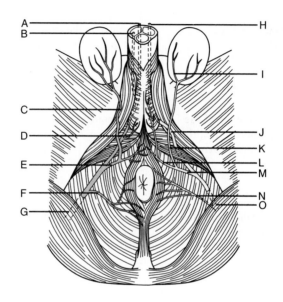

A = Dorsal artery of the penis
B = Deep artery of the penis
C = Posterior scrotal artery
D = Artery of the bulb of the penis
E = Perineal artery
F = Inferior rectal artery
G = Internal pudendal artery
H = Dorsal nerve of the penis
I = Posterior scrotal nerve
J = Dorsal nerve of the penis
K = Superficial perineal branch
L = Deep perineal branch
M = Perineal nerve
N = Inferior rectal nerve
O = Pudendal nerve

The pudendal nerve carries which:

    Sensory fibers?   Those that supply the external genitalia and perianal region

    Motor fibers?   Those that supply the external anal sphincter and muscles associated with ejaculation

Which branch of the pudendal nerve supplies the external anal sphincter?   The inferior rectal nerve

AUTONOMIC INNERVATION

Which nerves contribute to the inferior hypogastric plexus?

1. Parasympathetics, via the pelvic splanchnic nerves (derived from spinal segments S2–S4)
2. Sympathetics from the superior hypogastric plexus and pelvic sympathetic trunk

Afferent fibers sensing bladder fullness are carried by which nerves?   The pelvic splanchnic nerves

Visceral efferents to the detrusor muscle and internal sphincter of the bladder are carried by which nerves?   Parasympathetics

Which nerves supply motor innervation to the smooth muscle of the prostate, the seminal vesicle, the ejaculatory ducts, and the ductus deferens (vas deferens)?   Sympathetics

Which autonomic nerves coordinate:

    Erection?   Parasympathetics

    Ejaculation?   Sympathetics

PELVIC VISCERA

MALE PELVIC VISCERA

Identify the labeled structures on the following figure of the male pelvic viscera:

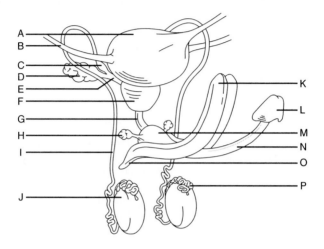

A = Bladder
B = Ureter
C = Ampulla of the ductus deferens
D = Seminal vesicle
E = Ejaculatory duct
F = Prostate gland
G = Urethra (membranous portion)
H = Bulbourethral gland (of Cowper)
I = Ductus deferens (vas deferens)
J = Testis
K = Corpus cavernosum
L = Glans penis
M = Bulb of the penis
N = Corpus spongiosum
O = Crus of the penis
P = Epididymis

Which ducts merge to form the ejaculatory ducts?	The duct of the seminal vesicle and the ductus deferens
Where does the ductus deferens enter the pelvic cavity?	At the deep inguinal ring, lateral to the inferior epigastric artery

How does the ductus deferens travel relative to the ureter?

The ductus deferens travels anterior to the ureter (as the ureter enters the bladder) to join with the duct of the seminal vesicle.

Where do the following structures open into the urethra:

Ejaculatory ducts?

At the seminal colliculus, lateral and inferior to the prostatic utricle (a remnant of the müllerian duct)

Bulbourethral glands (of Cowper)?

At the spongy urethra

Prostatic ducts?

At the prostatic sinuses (grooves alongside the urethral crest)

FEMALE PELVIC VISCERA

Identify the labeled structures on the following figure of the female pelvic viscera:

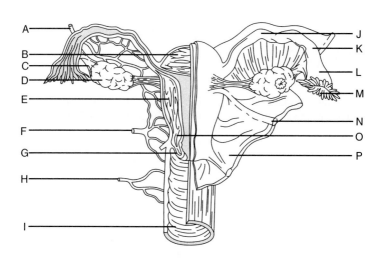

A = Ovarian artery
B = Fundus of the uterus
C = Ovarian ligament
D = Ovary
E = Body of the uterus
F = Uterine artery
G = Cervix

H = Vaginal artery
I = Vagina
J = Fallopian (uterine) tube
K = Ampulla of the fallopian tube
L = Infundibulum of the fallopian tube
M= Fimbriae
N = Round ligament of the uterus
O = Isthmus of the uterus
P = Broad ligament

What is the name of the recess between the cervix and the wall of the vagina?

The fornix

What are the three components of the fornix?

The lateral, anterior, and posterior components

What are the four parts of the uterus?

1. Cervix (external os)
2. Isthmus (internal os)
3. Body
4. Fundus

Name the three layers that comprise the wall of the body of the uterus.

1. Perimetrium
2. Myometrium
3. Endometrium

What is the normal relation of the uterus to the bladder?

The uterus normally overlies the bladder with the fundus positioned anteriorly, at roughly a right angle to the cervix

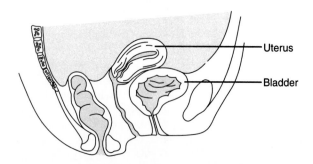

What is the normal position of the uterus?

Anteflexed and anteverted

What is the:

Vesicouterine pouch?

The intraperitoneal space between the bladder and the uterus

Pouch of Douglas (rectouterine pouch)?

The intraperitoneal space between the uterus and the rectum

What are the five parts of the fallopian tube?

1. Infundibulum
2. Ampulla
3. Isthmus
4. Intramural region
5. Fimbriae

Identify the three mesenteries of the ovary and fallopian tube:

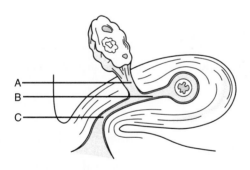

A = Mesovarium
B = Mesosalpinx
C = Mesometrium

Which ligament runs antero-inferiorly from the uterus within the broad ligament?

The round ligament of the uterus

Where do the ends of the round ligament of the uterus attach?

One end attaches to the anterolateral fundic region of the uterus; the other attaches in the labium majus.

Which two ligamentous portions of the broad ligament connect to the ovary?	The suspensory ligament of the ovary and the ovarian ligament proper
How does the suspensory ligament of the ovary differ from the ovarian ligament proper?	The suspensory ligament of the ovary is lateral to the ovary and contains the ovarian vessels, while the ovarian ligament proper is medial to the ovary and connects the ovary and the uterus.
Which ligament attaches the ovary to the wall of the pelvis?	The suspensory ligament of the ovary
Where do the ovarian veins drain to?	The right ovarian vein drains to the inferior vena cava and the left drains to the renal vein.

RECTUM AND ANAL CANAL

How long is the adult rectum?	Approximately 12 centimeters
Describe the peritoneal coverings of the rectum.	The superior third is covered anteriorly and laterally, the middle third is covered only anteriorly, and the inferior third has no peritoneal covering owing to peritoneal reflection onto the bladder (in men) or the vagina and uterus (in women).
Which arteries supply the rectum?	1. The superior rectal artery, the direct continuation of the inferior mesenteric artery 2. The middle rectal artery, a branch of the internal iliac artery 3. The inferior rectal artery, a branch of the internal pudendal artery

Identify the arterial supply of the rectum and associated structures on the following figure:

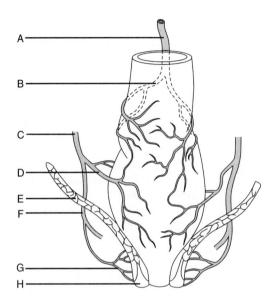

A = Superior rectal artery
B = Right branch of the superior rectal artery
C = Internal iliac artery
D = Middle rectal artery
E = Levator ani muscle group
F = Internal pudendal artery
G = Inferior rectal artery
H = External anal sphincter

Which veins drain the rectum?

The inferior and middle rectal veins (part of the systemic venous system) and the superior rectal vein (part of the portal venous system)

What is the dentate (pectinate) line?

The division between the superior and inferior anal canal

Where is the dentate line located?

Approximately two thirds of the way down the anal canal

What transitions occur at the dentate line?

1. Columnar or cuboidal epithelium above the dentate line changes to stratified squamous epithelium below it.
2. Processes above the dentate line (i.e., hemorrhoids) are painless; those below are painful.
3. Above the dentate line, visceral innervation is supplied via the hypogastric plexus; below the dentate line, somatic innervation is supplied via the pudendal nerve.
4. Venous drainage above the dentate line is via the portal venous system; below the dentate line, venous drainage is to the systemic venous system.
5. Lymphatic drainage above the dentate line is to the internal iliac nodes; below the dentate line, lymphatic drainage is to the superficial inguinal lymph nodes.

What are anorectal columns?

Longitudinal folds that extend from the anorectal junction to the dentate line

What are anal sinuses?

Pouch-like recesses between the anorectal column; the glands of the anal canal open into the anal sinuses

PERINEUM

What is the superior boundary of the perineum?

The pelvic floor

Which layer of abdominal fascia is Colles' fascia (i.e., the superficial perineal fascia) continuous with?

The membranous layer of the superficial abdominal fascia (i.e., Scarpa's fascia)

What else is Colles' fascia continuous with?

The tunica dartos in the scrotum and the superficial fascia of the penis

If you draw an imaginary line between the ischial tuberosities, which two anatomic regions are created?

1. Anal triangle
2. Urogenital triangle

What are the three main components of the anal triangle?

1. Anal canal
2. External anal sphincter
3. Ischioanal fossae

Describe the location and shape of the ischioanal fossae.

The ischioanal fossae are wedge-shaped areas located on each side of the anus.

What are the three major borders of the ischioanal fossae?

The base of the wedge is formed by the skin overlying the anal triangle on each side of the anus. The lateral walls are formed by the obturator internus muscle and fascia. The medial walls are formed by the levator ani muscle group (i.e., the pelvic diaphragm).

Which two nerves traverse the ischioanal fossa?

1. Inferior rectal nerve
2. Perineal branch of the femoral cutaneous nerve

What else does the ischioanal fossa contain?

Ischiorectal fat and the inferior rectal artery and vein

What are the three major parts of the external anal sphincter?

Subcutaneous, superficial, and deep

With which muscle does the deep part of the external anal sphincter blend?

The puborectalis muscle

Which nerve innervates the external anal sphincter?

The inferior rectal nerve

How does control of the external anal sphincter differ from that of the internal anal sphincter?

Control of the external anal sphincter is voluntary, whereas control of the internal anal sphincter is involuntary.

Which sphincter must relax for defecation to occur?

Both

What structure is at the apex of the urogenital triangle?

The symphysis pubis (i.e., imaginary lines connecting the symphysis pubis to each ischial tuberosity form the two sides of the triangle, and an imaginary line between the ischial tuberosities forms the bottom)

What is the deep perineal space?	The space enclosed by the inferior and superior fasciae of the urogenital diaphragm—these layers are continuous with each other anteriorly (near the symphysis pubis) and posteriorly, and attach to the pubic rami laterally

MALE PERINEUM

Penis

What are the three anatomic divisions of the penis, and which structures comprise each?	1. Root of the penis, comprised of the two crura and the bulb of the penis 2. Body of the penis, comprised of the corpus spongiosum and the two corpora cavernosa 3. Glans of the penis, comprised of the terminal corpus spongiosum
Where do the ischiocavernosus muscles:	
Originate?	The ischial tuberosities
Insert?	Penile crura
Where does the bulbospongiosus muscle:	
Originate?	From the perineal body
Insert?	The bulb, dorsum, and side of the penis
What is the action of the ischiocavernosus and bulbospongiosus muscles?	They retard venous return, thereby permitting erection to occur, and facilitate the expulsion of ejaculate at the time of ejaculation
What is the innervation of the ischiocavernosus and bulbospongiosus muscles?	The perineal branch of the pudendal nerve
What arteries supply the penis?	The dorsal and deep arteries of the penis, branches of internal pudendal artery
Between which layers of the penis are the dorsal arteries located?	Between the tunica albuginea (a fascial layer that envelops the corpus cavernosum and corpus spongiosum) and Buck's (deep) fascia

Where are the deep arteries of the penis located?

Within the corpora cavernosa

Where is the deep dorsal vein of the penis located?

Between the tunica albuginea and Buck's (deep) fascia (like the dorsal arteries)

What are the three anatomic divisions of the male urethra?

1. Prostatic urethra
2. Membranous urethra
3. Spongy urethra

Identify the labeled structures on the following figure of the penis:

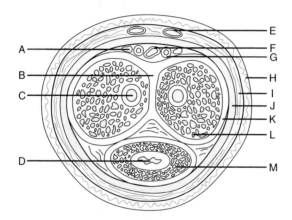

A = Dorsal nerve of the penis
B = Septum penis
C = Deep artery of the penis
D = Urethra
E = Superficial dorsal vein
F = Deep dorsal vein
G = Dorsal artery of the penis
H = Skin
I = Superficial fascia
J = Buck's (deep) fascia
K = Tunica albuginea
L = Corpus cavernosum
M = Corpus spongiosum

Scrotum and testes

What is the nature and action of the tunica dartos?

The tunica dartos is mostly composed of smooth muscle and functions to regulate the temperature for spermiogenesis by retracting and relaxing the scrotum.

What is the name of the fascia underlying the scrotal skin?

The dartos fascia

What structure lies directly internal to the internal spermatic fascia?

The tunica vaginalis

What are the two layers of the tunica vaginalis?

A visceral layer, which covers the testis, and a parietal layer

What lies directly internal to the visceral layer of the tunica vaginalis?

The tunica albuginea

Identify the lettered structures on the following diagram:

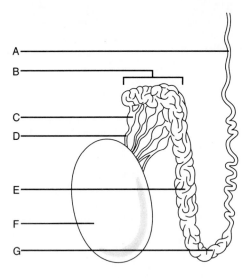

A = Ductus deferens
B = Head of the epididymis
C = Lobules of the epididymis
D = Efferent ductules of the testis
E = Body of the epididymis
F = Testis
G = Tail of the epididymis

What is the innervation of the scrotum?

The scrotum is innervated by the ilioinguinal, genitofemoral, perineal, and posterior femoral cutaneous nerves.

What is the lymphatic drainage of the:

   Scrotum?

Superficial inguinal lymph nodes

   Testis?

The aortic and retroperitoneal lymph nodes

Into which vein does the testicular vein drain on the:

   Left?

The left renal vein

   Right?

Inferior vena cava

FEMALE PERINEUM

Identify the labeled structures on the following figure of the external female genitalia:

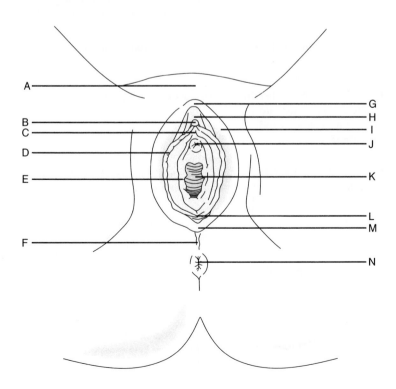

A = Mons pubis
B = Glans clitoris
C = Frenulum of the clitoris
D = Labium minus (labia minora, plural)
E = Hymen (ruptured)
F = Perineal body (site of the central perineal tendon)
G = Anterior labial commissure
H = Prepuce (hood of the clitoris)
I = Labium majus (labia majora, plural)
J = External urethral orifice
K = Vaginal orifice
L = Frenulum of the labia minora
M = Posterior labial commissure
N = Anus

POWER REVIEW

What is the difference between the pelvis minor (true pelvis) and the pelvis major (false pelvis)?	The pelvis major (false pelvis) is the wide portion of the bony pelvis that is located above the pelvic brim and is technically part of the abdomen. The pelvis minor (true pelvis) is the portion of the pelvis bounded by the pelvic inlet and outlet.
Which ligament creates the sciatic foramen?	The sacrotuberous ligament
Which ligament divides the sciatic foramen into the greater and lesser sciatic foramina?	The sacrospinous ligament
Where is the pudendal canal located?	On the medial border of the obturator internus muscle
What structure forms the pudendal canal?	The obturator internus (parietal pelvic) fascia
What is the origin of the levator ani muscle group called?	The tendinous arch
Which three muscles comprise the levator ani group?	1. Iliococcygeus muscle 2. Pubococcygeus muscle 3. Puborectalis muscle

From which artery does the testicular or ovarian artery arise on the:

   Left?   The aorta

   Right?   The aorta

Into which vein does the testicular or ovarian vein drain on the:

   Left?   The left renal vein

   Right?   Directly into the inferior vena cava

What are three branches of the posterior internal iliac artery?

1. Iliolumbar artery
2. Lateral sacral artery
3. Superior gluteal artery

Which arteries supply the rectum?

1. The superior rectal artery, the direct continuation of the inferior mesenteric artery
2. The middle rectal artery, a branch of the internal iliac artery
3. The inferior rectal artery, a branch of the internal pudendal artery

Which veins drain the rectum?

The inferior, middle, and superior rectal veins

Which rectal veins are:

   Systemic?   The inferior and middle rectal veins

   Portal?   The superior rectal vein

What is the origin and path of the pudendal nerve in the pelvis?

The pudendal nerve, a branch of the sacral plexus, accesses the gluteal region by exiting the pelvis through the greater sciatic foramen; it then reenters the pelvis through the lesser sciatic foramen to travel in the pudendal canal

Which three nerves arise from the pudendal nerve?

1. Inferior rectal nerve
2. Perineal nerve
3. Dorsal nerve of the penis (in men) or clitoris (in women)

Which surface of the bladder is covered by peritoneum?

The superior surface

Which three layers comprise the wall of the body of the uterus?

1. Perimetrium
2. Myometrium
3. Endometrium

What are the five parts of the fallopian tube?

1. Infundibulum
2. Ampulla
3. Isthmus
4. Intramural region
5. Fimbriae

11 ___

The Lower
Extremity

BONES

What are the four regions of the lower limb, and which bones are found in each region?

1. Hip: Ilium, ischium, and pubis
2. Thigh: Femur and patella
3. Leg: Tibia and fibula
4. Foot: Tarsal bones, metatarsal bones, and phalanges

THIGH

Identify the labeled structures on the following figure of the hip and thigh:

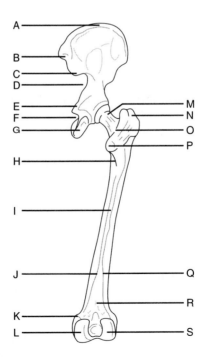

A = Iliac crest
B = Posterior superior iliac spine
C = Posterior inferior iliac spine
D = Greater sciatic notch
E = Ischial spine
F = Lesser sciatic notch
G = Ischial tuberosity
H = Pectineal line
I = Linea aspera
J = Medial supracondylar line
K = Adductor tubercle
L = Medial femoral condyle
M = Neck of the femur
N = Greater trochanter of the femur
O = Intertrochanteric crest
P = Lesser trochanter of the femur
Q = Lateral supracondylar line
R = Popliteal surface
S = Lateral femoral condyle

Femur

Describe the proximal end of the femur.

It consists of a head, neck, and the greater and lesser trochanters.

What structure separates the greater trochanter of the femur from the skin?

The trochanteric bursa

What structure attaches to the intertrochanteric line?

The iliofemoral ligament, a substantial ligament that contributes to the capsule of the hip joint

What is the intertrochanteric crest?

A prominent ridge that unites the two trochanters posteriorly

What is the quadrate tubercle?

A prominence on the intertrochanteric crest where the quadratus femoris muscle attaches

What is the fovea capitis?

A pit located roughly in the center of the femoral head, where the ligament of the femoral head attaches

What is the linea aspera?

A ridge that runs down the posterior midline of the femoral shaft

What becomes of the lateral and medial lips of the linea aspera?	The lateral lip becomes the gluteal tuberosity (the insertion site of the gluteus maximus muscle) and the medial lip continues as the spiral line to join the iliotibial line.
What connects the lesser trochanter with the linea aspera?	The pectineal line
Which structure attaches to the pectineal line?	The tendon of the pectineus muscle
Describe the distal end of the femur.	The distal end of the femur has medial and lateral condyles, which articulate with the tibia and patella to form the knee joint.

Describe the articular surfaces of the distal femur:

Anteriorly	The medial and lateral femoral condyles articulate with the patella.
Posteriorly	The medial and lateral condyles of the femur articulate with the medial and lateral tibial condyles of the tibia.

Patella

What type of bone is the patella?	A sesamoid bone (the largest in the body)
Where is the patella located?	Within the tendon of the quadriceps femoris muscle
Name two functions of the patella.	1. Attaches the quadriceps femoris tendon to the tibial tuberosity via the patellar ligament 2. Increases the power of the quadriceps femoris muscle by increasing its leverage

LEG

Identify the labeled land-marks on the following anterior view of the leg:

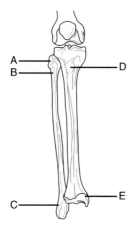

A = Head of the fibula
B = Neck of the fibula
C = Lateral malleolus (of the fibula)
D = Tibial tuberosity
E = Medial malleolus (of the tibia)

What is the name of the well-formed ridge on the posterior tibial surface?

The soleal line

Is the tibial "head" proximal or distal?

Proximal

Which process extends inferomedially from the distal end of the tibia?

The medial malleolus (of the tibia)

With which bone of the foot does the tibia articulate?

The talus (one of the seven tarsal bones)

FOOT

Identify the labeled structures on the following posterior view of the leg and the plantar aspect of the foot:

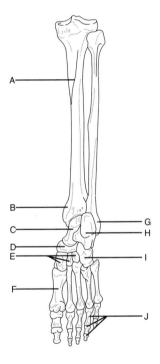

A = Soleal line (of the tibia)
B = Medial malleolus (of the tibia)
C = Talus
D = Navicular bone
E = Cuneiform bones
F = Metatarsal bone (1 of 5)
G = Lateral malleolus (of the fibula)
H = Calcaneus
I = Cuboid bone
J = Phalanges (3 of 14)

**Identify the labeled
structures on the following
lateral view of the foot:**

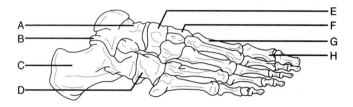

A = Talus
B = Lateral tubercle
C = Calcaneus
D = Cuboid bone
E = Navicular bone
F = Cuneiform bone (1 of 3)
G = Metatarsal bone (1 of 5)
H = Phalanx (1 of 14)

Name the seven tarsal bones.	1. Talus 2. Calcaneus 3. Cuboid bone 4. Navicular bone 5. Cuneiform bones (3)
What is the largest and most posterior tarsal bone?	The calcaneus
What structure inserts into the posterior surface of the calcaneus?	The tendo calcaneus (Achilles tendon)
The talus articulates with which two tarsal bones?	The navicular bone and the calcaneus
The calcaneus articulates with which two tarsal bones?	The talus and the cuboid bone
The navicular bone articulates with which five tarsal bones?	The talus, the cuboid bone, and the three cuneiform bones
Name the four intertarsal joints.	1. Talocalcaneonavicular 2. Talocalcaneal 3. Calcaneocuboid 4. Cuneonavicular

Which movements occur around the intertarsal joints?	Inversion and eversion
What structure passes through the large groove on the underlying surface of the cuboid bone?	The peroneus longus tendon
Describe the articulation of the metatarsal bones with the tarsal bones.	The first three metatarsal bones articulate with the cuneiform bones. The fourth and fifth metatarsal bones articulate with the cuboid bone (i.e., the cuboid bone lies laterally).
Which metatarsal bone has two articular facets for articulation with sesamoid bones?	The first metatarsal bone
Which muscle inserts into a tuberosity at the base of the fifth metatarsal bone?	The peroneus brevis muscle

GLUTEAL REGION

Name the nine muscles of the gluteal region.	1. Gluteus maximus muscle 2. Gluteus minimus muscle 3. Gluteus medius muscle 4. Tensor fasciae latae muscle 5. Piriformis muscle 6. Obturator internus muscle 7. Gemellus inferior muscle 8. Gemellus superior muscle 9. Quadratus femoris muscle
Which artery supplies the:	
Gluteus maximus muscle?	The inferior gluteal artery
Gluteus medius and minimus muscles?	The superior gluteal artery

GLUTEUS MAXIMUS MUSCLE

What is the gluteus maximus muscle's "claim to fame?"	It is the largest single muscle in the body.

Name the three bursae that separate the gluteus maximus muscle from the underlying structures.	1. Trochanteric bursa 2. Gluteofemoral bursa 3. Ischial bursa
What is the origin of the gluteus maximus muscle?	The ilium, sacrum, coccyx, and sacrotuberous ligament
Insertion?	The gluteal tuberosity (i.e., the lateral lip of the linea aspera of the femur) and the iliotibial tract (the portion of the fascia lata that extends from the iliac crest to the tibia)
Innervation?	The inferior gluteal nerve
Action?	Chief extensor of the thigh and a lateral rotator of the hip joint

GLUTEUS MINIMUS MUSCLE

Origin?	The ilium, between the anterior and inferior gluteal lines
Insertion?	The greater trochanter of the femur
Innervation?	The superior gluteal nerve
Action?	Abducts and rotates the thigh medially

GLUTEUS MEDIUS MUSCLE

Origin?	The ilium, between the anterior and posterior gluteal lines
Insertion?	The greater trochanter of the femur
Innervation?	The superior gluteal nerve
Action?	Abducts the hip joint and, along with the gluteus minimus muscle, maintains stability of the pelvis during ambulation

TENSOR FASCIAE LATAE MUSCLE

Origin?	The anterior superior iliac spine and the external lip of the iliac crest

Insertion?   The lateral tibial condyle (via the iliotibial tract)

Innervation?   The superior gluteal nerve

Action?   Abducts, medially rotates, and flexes the thigh

PIRIFORMIS MUSCLE

Origin?   The pelvic surface of the sacrum and the sacrotuberous ligament

Insertion?   The upper end of the greater trochanter of the femur

Innervation?   The sacral nerve

Action?   Rotates the thigh laterally

OBTURATOR INTERNUS MUSCLE

Origin?   The ischiopubic rami and the obturator membrane

Insertion?   The greater trochanter of the femur

Innervation?   The nerve to the obturator internus

Action?   Abducts and laterally rotates the thigh

GEMELLI MUSCLES

Origin?   **Gemellus superior:** The ischial spine
                             Gemellus inferior: The ischial tuberosity

Insertion?   The obturator internus tendon

Innervation?   **Gemellus superior:** The nerve to the obturator internus
                             Gemellus inferior: The nerve to the quadrate femoris

Action?   Abducts and rotates the thigh

QUADRATUS FEMORIS MUSCLE

Origin?	The ischial tuberosity
Insertion?	The quadrate tubercle of the intertrochanteric crest
Innervation?	The nerve to the quadratus femoris
Action?	Rotates the thigh laterally

HIP REGION

What type of joint is the hip joint?	A synovial ball-and socket joint
Which bones articulate at the hip joint?	The head of the femur and the hip bone (at the acetabulum)
Which muscle is the major flexor at the hip joint?	The iliopsoas muscle, one of the anterior thigh muscles
Name the five ligaments that are associated with the hip joint.	1. Iliofemoral ligament 2. Ischiofemoral ligament 3. Pubofemoral ligament 4. Transverse acetabular ligament 5. Ligament capitis femoris
Which of these ligaments is the largest and most important clinically?	The iliofemoral ligament
Which ligament is located at the base of the hip joint?	The transverse acetabular ligament

THIGH

What is the thigh?	The limb segment between the hip and knee joints

MUSCLES

Name the three muscle compartments of the thigh.	Posterior, medial, and anterior

Posterior thigh muscles

List the four muscles of the posterior thigh compartment.	1. Semimembranous muscle 2. Semitendinosus muscle

3. Biceps femoris muscle (long and short heads)
4. Adductor magnus muscle (hamstring part)

What are the "hamstring" muscles?

The semimembranosus muscle, the semitendinosus muscle, the long head of the biceps femoris muscle, and the adductor magnus muscle (hamstring part)

Identify the muscles and related structures on the following figure of the gluteal region and posterior thigh:

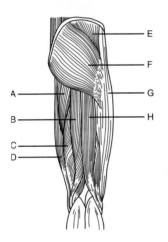

A = Adductor magnus muscle (hamstring part)
B = Semitendinosus muscle
C = Semimembranosus muscle
D = Gracilis muscle
E = Gluteus medius muscle
F = Gluteus maximus muscle
G = Iliotibial tract
H = Biceps femoris muscle

What is the common origin of the posterior thigh muscles?

The ischial tuberosity

What is the common innervation of the posterior thigh muscles?

The tibial division of the sciatic nerve

Semimembranosus muscle

Origin?

The ischial tuberosity

Insertion?

The posterior part of the medial tibial condyle

Innervation?

The sciatic nerve (tibial division)

Action?	Extends the thigh and flexes and medially rotates the leg

Semitendinosus muscle

Origin?	The ischial tuberosity
Insertion?	The superomedial tibia
Innervation?	The sciatic nerve (tibial division)
Action?	Extends the thigh and flexes and medially rotates the leg

Biceps femoris muscle

Origin?	**Long head:** The ischial tuberosity **Short head:** The lateral lip of the linea aspera and the lateral supracondylar line of the femur
Insertion?	The head of the fibula (lateral side)
Innervation?	The sciatic nerve (the tibial division innervates the long head and the common peroneal division innervates the short head)
Action?	Flexes and laterally rotates the leg and extends the thigh

Adductor magnus muscle (hamstring part)

Origin?	The ischial tuberosity
Insertion?	The adductor tubercle of the medial epicondyle of the femur
Innervation?	The sciatic nerve (tibial division)
Action?	Adducts and extends the thigh

Medial thigh muscles

List the six muscles of the medial thigh compartment.	1. Pectineus muscle 2. Adductor longus muscle 3. Adductor magnus muscle (adductor part) 4. Adductor brevis muscle 5. Gracilis muscle 6. Obturator externus muscle

Where do the three "adductors" insert?	On the linea aspera of the femur
Which medial thigh muscles other than the three "adductors" contribute to thigh adduction?	The gracilis muscle and the pectineus muscle
Which of these two muscles is medial?	The gracilis muscle (the pectineus muscle is lateral)
Of the five muscles in the medial thigh group that contribute to adduction, which can be removed without noticeable loss of function?	The gracilis muscle (often transplanted with its nerves and blood supply to replace a damaged muscle)

Pectineus muscle

Origin?	The pecten pubis (pectineal line of the pubis)
Insertion?	The pectineal line of the femur
Innervation?	The femoral nerve
Action?	Adducts and flexes the thigh

Adductor longus muscle

Origin?	The body of the pubis (inferior to the pubic crest)
Insertion?	The middle third of the linea aspera of the femur
Innervation?	The obturator nerve
Action?	Adducts the thigh

Adductor magnus muscle (adductor part)

Origin?	The pubic and ischial rami
Insertion?	The gluteal tuberosity, the medial linea aspera, and the supracondylar line of the femur

Innervation?   The obturator nerve

Action?   Adducts and flexes the thigh

Adductor brevis muscle

Origin?   The body and inferior ramus of the pubis

Insertion?   The pectineal line and the proximal part
                                    of the linea aspera of the femur

Innervation?   The obturator nerve

Action?   Adducts and flexes the thigh

Gracilis muscle

Origin?   The body and inferior ramus of the pubis

Insertion?   The superomedial tibia

Innervation?   The obturator nerve

Action?   Adducts the thigh and flexes and medially
                                    rotates the leg

Obturator externus muscle

Origin?   The obturator membrane and the
                                    margins of the obturator foramen

Insertion?   The trochanteric fossa of the femur

Innervation?   The obturator nerve

Action?   Laterally rotates the thigh and steadies
                                    the femoral head in the acetabulum

Anterior thigh muscles

List the three muscles of the   1. Iliopsoas muscle
anterior compartment of   2. Sartorius muscle
the thigh.   3. Quadriceps femoris muscle

Iliopsoas muscle

Which two muscles com-   The iliacus and the psoas major muscles
prise the iliopsoas muscle?

Iliacus muscle

Origin?   The iliac crest, iliac fossa, sacral ala, and
                                    the anterior sacroiliac ligament

Insertion?	Inferior to the lesser trochanter of the femur, via the iliopsoas tendon
Innervation?	The femoral nerve
Action?	Flexes the hip and stabilizes the hip joint in conjunction with the psoas major muscle

Psoas major muscle

Origin?	The bodies and intervertebral disks of vertebrae T12–L5
Insertion?	The lesser trochanter of the femur (via the iliopsoas tendon)
Innervation?	The ventral primary rami of spinal cord levels L1–L3
Action?	Flexes the hip and stabilizes the hip joint in conjunction with the iliacus muscle

Sartorius muscle

What is the sartorius muscle's claim to fame?	It is the longest muscle in the body.
What is the sartorius muscle's origin?	The anterior superior iliac spine and from the ilial notch, below the anterior superior iliac spine
Insertion?	The superomedial tibia
Innervation?	The femoral nerve
Action?	Flexes, abducts, and laterally rotates the thigh at the hip and flexes the knee

Quadriceps femoris muscle

Which four muscles contribute to the quadriceps femoris muscle?	1. Rectus femoris muscle 2. Vastus lateralis muscle 3. Vastus medialis muscle 4. Vastus intermedius muscle

State where each of the four muscles that contributes to the quadriceps femoris muscle originates:

Rectus femoris muscle?	The anterior inferior iliac spine and the ilial ala (superior to the acetabulum)
The vastus lateralis muscle?	The greater trochanter and lateral lip of the linea aspera of the femur
Vastus medialis muscle?	The intertrochanteric line and the medial lip of the linea aspera of the femur
Vastus intermedius muscle?	The anterior and lateral surfaces of the body of the femur

Where does the quadriceps femoris muscle insert?

The quadriceps tendon inserts into the superior border of the patella and continues as the patellar ligament to insert on the tibial tuberosity.

Which nerve innervates the quadriceps femoris muscle?

The femoral nerve

What is the action of the quadriceps femoris muscle?

It extends the leg at the knee joint.

FASCIAE

Name the deep fascia of the thigh.

Fascia lata

What is the name of the thick, strong, lateral portion of the fascia lata in the thigh?

The iliotibial tract

Where does the iliotibial tract originate?

On the tubercle of the iliac crest

Where does it insert?

On the lateral condyle (of the tibia)

What is the function of the iliotibial tract?

It acts as tendon for the tensor fasciae latae muscle and contributes to the tendon of the gluteus maximus muscle, thereby steadying the trunk on the thighs.

What structures are housed within the superficial fascia of the thigh (tela subcutanea)?

The cutaneous vessels and the superficial inguinal lymph nodes

VASCULATURE

Arteries

Identify the labeled arteries of the lower limb on the following figure:

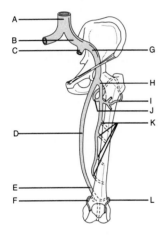

A = Aorta
B = Common iliac artery
C = Internal iliac artery
D = Femoral artery
E = Popliteal artery
F = Superior medial geniculate artery
G = External iliac artery
H = Deep femoral artery (profunda femoris)
I = Lateral circumflex femoral artery
J = Medial circumflex femoral artery
K = Perforating arteries
L = Superior lateral geniculate artery

Which arteries represent the primary arterial supply to the thigh?

The femoral and deep femoral arteries

Describe the course of the deep femoral artery.

The deep femoral artery arises from, and runs lateral to, the femoral artery before passing posterior to it and the femoral vein. It passes between the pectineus and adductor longus muscles and descends posterior to the latter muscle, giving off perforating branches that supply the adductor magnus muscle and the posterior thigh muscles.

Which branches of the deep femoral artery supply the head and neck of the femur and the muscles on the lateral side of the thigh?

The medial and lateral circumflex femoral arteries

Veins

**Identify the labeled veins
of the lower extremity on
the following figure:**

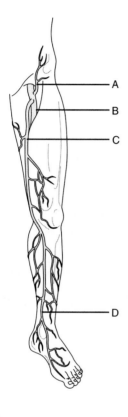

A = Superficial circumflex iliac vein
B = Femoral vein
C = Great saphenous vein
D = Great saphenous vein

**What are the two most
important tributaries of the
femoral vein?**

The great saphenous vein and the deep
femoral vein

**What is the longest vein in
the body?**

The great saphenous vein

**Name the five major tribu-
taries to the great saphenous
vein in the thigh.**

1. Superficial circumflex iliac vein
2. Superficial epigastric vein
3. External pudendal vein
4. Lateral femoral cutaneous vein
5. Anterior femoral cutaneous vein

What other vein may drain into the great saphenous vein, but is not present in all individuals?	The accessory saphenous vein, a major communicating vein between the great and small saphenous veins
What is the function of the perforating (anastomotic) veins of the lower extremity?	They drain the superficial venous system into the deep venous system.

INNERVATION

Which cutaneous nerve innervates most of the:

Anterior and medial thigh?	The femoral nerve
Posterior thigh and popliteal fossa?	The posterior femoral cutaneous nerve (a branch of the sacral plexus)
Cutaneous innervation to much of the lateral thigh is via which nerve?	The lateral femoral cutaneous nerve (a branch of the lumbar plexus)
Which three nerves provide cutaneous innervation to the superomedial thigh?	1. Obturator nerve (cutaneous branch) 2. Ilioinguinal nerve 3. Genitofemoral nerve
How does the obturator nerve enter the thigh?	Via the obturator foramen
What is the largest nerve in the body?	The sciatic nerve
How does the sciatic nerve enter the thigh?	Via the greater sciatic notch, below the piriformis muscle
Nerves from which vertebral levels supply fibers to the sciatic nerve?	L4–S3

FEMORAL TRIANGLE

What are the boundaries of the femoral triangle:

Superiorly?	The inguinal ligament
Medially?	The medial border of the adductor longus muscle

Laterally?	The medial border of the sartorius muscle
What forms the floor of the femoral triangle?	The adductor longus muscle, the pectineus muscle, and the iliopsoas muscle
The roof?	The fascia lata

Name four structures that are contained with the femoral triangle.

1. The femoral nerve and its branches
2. The femoral artery and its branches
3. The femoral vein and its tributaries
4. The lymphatics that drain the lower limb (contained within an "empty" space filled with fat and connective tissue)

To remember the structures contained within the femoral triangle (from lateral to medial), think **NAVEL:**

Nerve

Artery

Vein

Empty space containing **L**ymphatics

Describe the location of the femoral artery in the femoral triangle.	The femoral artery is located 2–3 cm inferior to the midpoint of the inguinal ligament (just medial to the femoral nerve).
What is the femoral sheath?	A fascial tube that surrounds the femoral artery, the femoral vein, and the empty space containing the lymphatics within the femoral triangle.
The femoral sheath is in continuity with which layer of anterior abdominal wall fascia?	The transversalis fascia

What are the three compartments of the femoral sheath and what is contained in each?

1. The lateral compartment (contains the femoral artery)
2. The intermediate compartment (contains the femoral vein)
3. The medial compartment, or femoral canal (the "empty" space containing the lymphatics)

Which structure in the femoral triangle is NOT contained within the femoral canal?	The femoral nerve

ADDUCTOR CANAL

What is the adductor canal?	A fascial tunnel in the thigh located deep to the sartorius muscle
What are the boundaries of the adductor canal:	
Laterally?	The vastus medialis muscle
Posteromedially?	The adductor longus and adductor magnus muscles
Anteriorly?	The sartorius muscle
What forms the roof of the adductor canal?	The sartorius muscle and the subsartorial fascia
What four structures pass through the adductor canal?	1. Femoral artery 2. Femoral vein 3. Saphenous nerve 4. Nerve to the vastus medialis

KNEE REGION

KNEE JOINT

Which bones comprise the knee joint?	The femur, tibia, and patella (note that the fibula does not articulate with the patella)
Identify the labeled structures on the following anterior view of the dissected knee joint:	

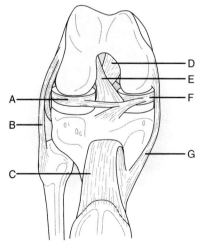

A = Lateral meniscus
B = Fibular (lateral) collateral ligament
C = Patellar ligament
D = Posterior cruciate ligament
E = Anterior cruciate ligament
F = Medial meniscus
G = Tibial (medial) collateral ligament

Identify the labeled structures on the following posterior view of the dissected knee joint:

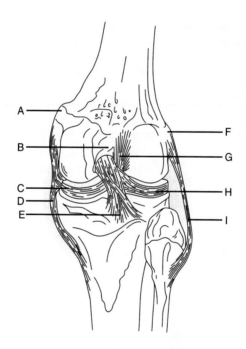

A = Medial femoral epicondyle
B = Intercondylar notch
C = Medial meniscus
D = Tibial (medial) collateral ligament
E = Posterior cruciate ligament
F = Lateral femoral epicondyle
G = Anterior cruciate ligament
H = Lateral meniscus
I = Fibular (lateral) collateral ligament

Which ligament runs from the lateral femoral epicondyle to the head of the fibula?

The fibular (lateral) collateral ligament

Which ligament runs from the medial femoral epicondyle to the medial surface of the tibial shaft?

The tibial (medial) collateral ligament

Which oblique ligament arises on the:

Posterior tibia and inserts on the medial femoral condyle?

The posterior cruciate ligament

Anterior tibia and inserts on the lateral femoral condyle?

The anterior cruciate ligament

Why are the anterior and posterior cruciate ligaments so named?

Because the two ligaments cross one another like an "X"

What is the patellar ligament?

The continuation of the quadriceps tendon from the apex of the patella to the tibial tuberosity

What are the major flexors at the knee joint?

The posterior thigh muscles (i.e., the semimembranosus, semitendinosus, biceps femoris, and hamstring portion of the adductor magnus muscles) and two of the posterior leg muscles (i.e., the gastrocnemius and plantaris muscles)

Which muscle is the prime extensor at the knee joint?

The quadriceps femoris muscle

What muscle provides for some rotation at the knee?

The popliteus muscle, a posterior leg muscle that courses from the lateral femoral condyle to the popliteal surface of the tibia

Why can the knee only rotate when it is in the fixed position?

Flexion at the knee allows for relaxation of the collateral ligaments.

Name the four bursae of the knee.	1. Suprapatellar bursa 2. Prepatellar bursa 3. Subcutaneous infrapatellar bursa 4. Deep infrapatellar bursa
Which is formed by a superior extension of the synovial membrane of the knee joint?	The suprapatellar bursa (a favorite site for joint aspiration)

POPLITEAL FOSSA

What is the popliteal fossa?	The diamond-shaped area behind the knee joint
What are the boundaries of the popliteal fossa:	
Posteriorly?	The skin and superficial fascia
Superomedially?	The semimembranosus and semitendinosus muscles
Superolaterally?	The biceps femoris muscle
Inferomedially?	The medial head of the gastrocnemius muscle
Inferolaterally?	The lateral head of the gastrocnemius muscle and the plantaris muscle
Anteriorly?	The posterior surface of the femur, the posterior capsule of the knee joint, and the posterior surface of the tibia superior to the soleus muscle
What does the popliteal fossa contain?	1. Fat 2. The popliteal artery and vein 3. The small saphenous vein 4. Lymphatics and the popliteal lymph nodes 5. The tibial nerve 6. The common peroneal nerve 7. The posterior femoral cutaneous nerve 8. The articular branch of the obturator nerve 9. The popliteus bursa

LEG

MUSCLES

Name the three muscle compartments of the leg.	Posterior, lateral, and anterior

Posterior leg muscles

Name the seven muscles of the posterior compartment of the leg.	1. Soleus muscle 2. Gastrocnemius muscle 3. Plantaris muscle 4. Popliteus muscle 5. Flexor digitorum longus muscle 6. Flexor hallucis longus muscle 7. Tibialis posterior muscle
Which of these muscles are superficial?	The soleus, gastrocnemius, and plantaris muscles
What is the name of the common tendon for the gastrocnemius and soleus muscles?	The tendo calcaneus (Achilles tendon)
What is unique about the tendo calcaneus?	It is the longest tendon in the body.
Where does the tendo calcaneus insert?	On the posterior surface of the calcaneus, at the tuber calcanei
Which nerve innervates all of the muscles of the posterior compartment of the leg?	The tibial nerve

Soleus muscle

Origin?	The posterior aspect of the head of the fibula and the soleal line on the tibia
Insertion?	The posterior surface of the calcaneus, via the tendo calcaneus
Action?	Plantarflexes the foot

Gastrocnemius muscle

How many heads does the gastrocnemius muscle have?	Two

What is the origin of the gastrocnemius muscle?	The lateral femoral condyle (lateral head) and the medial femoral condyle (medial head)
Insertion?	The posterior surface of the calcaneus, via the tendo calcaneus
Action?	Plantarflexes the foot, flexes the leg, and raises the heel during ambulation

Plantaris muscle

Where is the plantaris muscle located?	Deep to the lateral head of the gastrocnemius muscle (i.e., between the gastrocnemius and soleus muscles)
What is the origin of the plantaris muscle?	The lower lateral supracondylar line
Insertion?	The posterior surface of the calcaneus
Action?	Plantarflexes the foot, flexes the leg

Popliteus muscle

Origin?	The lateral condyle of the femur and the arcuate popliteal ligament
Insertion?	The posterior surface of the tibia, superior to the soleal line
Action?	Flexes and rotates the leg medially

Flexor digitorum longus muscle

Origin?	The posterior surface of the shaft of the tibia
Insertion?	The bases of the distal phalanges of the lateral four toes
Action?	Flexes the lateral four toes and plantarflexes the foot

Flexor hallucis longus muscle

Origin?	The posterior surface of the shaft of the fibula
Insertion?	The base of the distal phalanx of the great toe

Action?	Flexes the distal phalanx of the great toe

Tibialis posterior muscle

Origin?	The posterior interosseus membrane, the tibia and the fibula
Insertion?	The navicular, cuneiform, and cuboid bones (i.e., all of the tarsal bones except the talus) and the bases of the second, third, and fourth metatarsal bones
Action?	Plantarflexes and inverts the foot

Lateral leg muscles

Name the two lateral leg muscles.	1. Peroneus longus muscle 2. Peroneus brevis muscle
Which nerve innervates both lateral leg muscles?	The superficial peroneal nerve

Peroneus longus muscle

Origin?	The head and upper lateral surface of the fibula
Insertion?	The base of the first metatarsal bone and the first cuneiform bone
Action?	Everts and plantarflexes the foot

Peroneus brevis muscle

Origin?	The lower lateral side of the fibula and the intermuscular septa
Insertion?	A tuberosity at the base of the fifth metatarsal bone
Action?	Everts and plantarflexes the foot

Anterior leg muscles

Name the four muscles of the anterior compartment of the leg.	1. Tibialis anterior muscle 2. Extensor digitorum longus muscle 3. Extensor hallucis longus muscle 4. Peroneus tertius muscle
Which nerve innervates the anterior leg muscles?	The deep peroneal nerve

Tibialis anterior muscle

Origin?

The lateral tibial condyle and the interosseus membrane

Insertion?

The first cuneiform bone and the first metatarsal bone

Action?

Dorsiflexes and inverts the foot

Extensor digitorum longus muscle

Origin?

The lateral tibial condyle, the proximal two thirds of the fibula, and the interosseous membrane

Insertion?

The bases of the middle and distal phalanges of the lateral four toes

Action?

Extends the toes and dorsiflexes the ankle joint

Extensor hallucis longus muscle

Origin?

The middle half of the anterior surface of the fibula and the interosseous membrane

Insertion?

The base of the distal phalanx of the great toe

Action?

Extends the great toe and dorsiflexes and inverts the foot

Peroneus tertius muscle

Origin?

The distal third of the fibula and the interosseus membrane

Insertion?

The dorsum of the shaft of the fifth metatarsal bone

Action?

Dorsiflexes and everts the foot

VASCULATURE

Which artery lies on the posterior cruciate ligament?

The popliteal artery

The popliteal artery bifurcates into which two branches at the lower border of the popliteus muscle?

The anterior and posterior tibial arteries (the initial portion of the posterior tibial artery is occasionally referred to as the tibioperoneal trunk).

Describe the course of the anterior tibial artery.

The anterior tibial artery passes through an opening in the interosseus membrane medial to the fibula, and then becomes the dorsalis pedis artery at the ankle joint.

Describe the course of the posterior tibial artery.

After giving off the peroneal artery, the posterior tibial artery runs distally within the posterior deep compartment; passes behind the medial malleolus (of the tibia); and then terminates by dividing into the medial and lateral plantar arteries in the sole of the foot.

INNERVATION

The sciatic nerve ends at the popliteal fossa, branching into which two nerves?

The tibial nerve and the common peroneal nerve

Describe the course of the tibial nerve in the leg.

It runs between the superficial and deep posterior leg muscles, and then runs along the tibia.

The tibial nerve bifurcates into which two nerves at the lower border of the popliteus muscle?

The anterior and posterior tibial nerves

Describe the course of the common peroneal nerve in the leg.

It runs along the fibula.

What are the branches of the common peroneal nerve?

The superficial peroneal nerve and the deep peroneal nerve

Which muscles are innervated by the superficial peroneal nerve?

The lateral leg muscles (i.e., the peroneus longus and the peroneus brevis muscles)

Which muscles are innervated by the deep peroneal nerve?

The anterior leg muscles (i.e., the tibialis anterior, extensor hallucis longus, extensor digitorum longus, and peroneus tertius muscles) and the extensor digitorum brevis and extensor hallucis brevis muscles of the foot

ANKLE REGION

Which bones articulate at the ankle joint?	The tibia, fibula, and talus
Which movements occur at the ankle joint?	Dorsiflexion (toes up) and plantarflexion (toes down) are major movements. Rotation, abduction, and adduction are possible when the foot is plantarflexed.
Which joint provides for most of the foot's ability to invert or evert?	The talocalcaneonavicular joint
Which two muscles invert the foot (i.e., turn the sole inward)?	The tibialis anterior and the tibialis posterior muscles
Which three muscles evert the foot (i.e., turn the sole outward)?	The peroneus longus, peroneus brevis, and peroneus tertius muscles
Which muscles are responsible for dorsiflexion at the ankle?	The extensor hallucis longus and extensor digitorum longus muscles dorsiflex the foot. The tibialis anterior muscle dorsiflexes and inverts the foot.
Which muscles are responsible for plantarflexion at the ankle?	The flexor hallucis longus, flexor digitorum longus, tibialis posterior, gastrocnemius, soleus, and plantaris muscles
Describe the medial and lateral ligaments of the ankle joint.	The medial ligament is deltoid-shaped and is thicker and stronger than the lateral ligament. The lateral ligament consists of three bands (the anterior and posterior talofibular ligaments and the middle calcaneofibular band).
Which two tendons pass behind the lateral malleolus (of the fibula) at the ankle joint?	The tendons of the peroneus longus and peroneus brevis muscles
Which two structures hold these two tendons in place?	The superior and inferior peroneal retinacula

FOOT

DORSAL FOOT MUSCLES

What are the dorsal foot muscles?
The extensor hallucis brevis muscle and the extensor digitorum brevis muscle

What is the action of these muscles?
They extend the toes.

Where do these muscles originate?
On the dorsal surface of the calcaneus

Where does the extensor hallucis brevis muscle insert?
On the base of the proximal phalanx of the great toe

Where does the extensor digitorum brevis muscle insert?
The three tendons of the extensor digitorum brevis muscle attach to the corresponding tendons of the extensor digitorum longus muscle, which insert on the bases of the middle and distal phalanges of the second, third, and fourth digits

Which nerve innervates the dorsal muscles of the foot?
The deep peroneal nerve

PLANTAR FOOT MUSCLES

What is the function of the muscles of the sole of the foot?
They help maintain the arches of the feet and enable a person to stand, even on an uneven surface. Their actions are gross, not specific.

List the three muscles that form the first (superficial) layer of muscles in the sole of the foot.
1. Abductor hallucis muscle
2. Abductor digiti minimi muscle
3. Flexor digitorum brevis muscle

List the two muscles that form the second layer of muscles in the sole of the foot.
1. Quadratus plantae muscle
2. Lumbrical muscles (four)

List the three muscles that form the third layer of muscles in the sole of the foot.
1. Flexor hallucis brevis muscle
2. Adductor hallucis muscle
3. Flexor digiti minimi brevis muscle

List the two groups of muscles that form the fourth layer of muscles in the sole of the foot.	1. Plantar interossei (three) 2. Dorsal interossei (four)

POWER REVIEW

GLUTEAL REGION

Which three gluteal muscles originate from the external surface of the ilium?	The gluteus maximus, gluteus medius, and gluteus minimus muscles
What is the innervation of the:	
Gluteus maximus muscle?	The inferior gluteal nerve
Gluteus minimus and gluteus medius muscles?	The superior gluteal nerve
Which other gluteal muscle is supplied by the superior gluteal nerve?	The tensor fasciae latae
Which two gluteal muscles insert into the iliotibial tract?	The gluteus maximus muscle and the tensor fasciae latae muscle
What is the name of the square muscle that runs from the ischial tuberosity to the intertrochanteric crest of the femur?	The quadratus femoris muscle

HIP

Which muscle is the major flexor at the hip joint?	The iliopsoas muscle (a combination of the iliacus and psoas major muscles), one of the anterior thigh muscles
Which muscle is the chief extensor of the thigh at the hip joint?	The gluteus maximus muscle

THIGH

Name the contents of the femoral triangle.	**NAVEL** (from lateral to medial): femoral **N**erve femoral **A**rtery femoral **V**ein **E**mpty space containing **L**ymphatics

Which arteries represent the primary arterial supply to the thigh?	The femoral and deep femoral (profunda femoris) arteries
What is the fascia of the thigh called?	The fascia lata
The great saphenous vein empties into which vein?	The femoral vein
What are the contents of the adductor canal?	The femoral artery and vein, the saphenous nerve, and the nerve to the vastus medialis

Which muscles comprise the:

Posterior thigh muscles?	Semimembranosus, semitendinosus, biceps femoris, and adductor magnus (hamstring part) muscles
Medial thigh muscles?	Pectineus, adductor longus, adductor magnus (adductor part), adductor brevis, gracilis, and obturator externus muscles
Anterior thigh muscles?	Iliopsoas, sartorius, and quadriceps femoris muscles

Which nerve innervates the:

Posterior thigh muscles?	The tibial division of the sciatic nerve
Medial thigh muscles?	The obturator nerve
Anterior thigh muscles?	The femoral nerve

KNEE

What are the major flexors at the knee joint?	1. The semimembranosus, semitendinosus, biceps femoris, and adductor magnus (hamstring portion) muscles (i.e., the posterior thigh muscles) 2. The gastrocnemius and plantaris muscles (posterior leg muscles)

Which muscle is the prime extensor at the knee joint?	The quadriceps femoris muscle
The popliteal artery bifurcates into which two branches at the lower border of the popliteus muscle?	The anterior and posterior tibial arteries

LEG

Name the muscles of the three compartments of the leg.	Posterior leg muscles: Soleus, plantaris, gastrocnemius, popliteus, flexor digitorum longus, flexor hallucis longus, and tibialis posterior
	Lateral leg muscles: Peroneus longus and peroneus brevis
	Anterior leg muscles: Tibialis anterior, extensor digitorum longus, extensor hallucis longus, and peroneus tertius

ANKLE

At the ankle joint, the anterior tibial artery becomes which artery?	The dorsalis pedis artery

Index